THE 10-MINUTE GARDENER

How to Have a Veggie Garden and a Life—
85 Time-Saving Tactics to Be More Efficient
and Grow More Food in Less Time

CaliKim

COOL SPRINGS PRESS

Quarto.com

© 2025 Quarto Publishing Group USA Inc.
Text © 2025 CaliKim Garden and Home DIY Inc
Photography © 2025 CaliKim Garden and
Home DIY Inc

First Published in 2025 by Cool Springs Press,
an imprint of The Quarto Group,
100 Cummings Center, Suite 265-D,
Beverly, MA 01915, USA.
T (978) 282-9590   F (978) 283-2742

Cool Springs Press titles are also available at
discount for retail, wholesale, promotional, and
bulk purchase. For details, contact the Special
Sales Manager by email at specialsales@
quarto.com or by mail at The Quarto Group,
Attn: Special Sales Manager, 100 Cummings
Center, Suite 265-D, Beverly, MA 01915, USA.

29 28 27 26 25   1 2 3 4 5

ISBN: 978-0-7603-9186-0

Digital edition published in 2025
eISBN: 978-0-7603-9187-7

Library of Congress Cataloging-in-Publication
Data is available.

Design and Page Layout: Laura Shaw Design
Photography: CaliKim Garden and Home DIY Inc

Printed in China

Dedicated to Jerry, my amazing CameraGuy. Your unwavering support, love, and faith in me makes the magic happen. Thanks for helping me multiply my time many times over and adding so much joy and fun to our lives!

Contents

INTRODUCTION

You *Can* Have a Garden and a Life

Have you ever found yourself saying "I don't have time to grow a garden"? This is *the* number one reason why many aspiring gardeners don't take the plunge. The truth is that even with a hectic, on-the-go lifestyle, you *can* grow a stunning, thriving garden and still live life to the fullest. It's all about gardening smarter, not working harder, and this book is your guide to achieving just that.

Perhaps you've envisioned a thriving vegetable garden filled with fresh, organic produce and vibrant flowers that are buzzing with bees. Maybe you've pictured yourself strolling down lovely garden pathways amid beautiful cedar raised beds with a harvest basket in hand, filling it to the brim with an abundance of veggies to share with family and friends. But then reality hits. The demands of daily life—work, kids, chores, and more—threaten to squash your dream.

The constant juggling of tasks leaves little room for gardening. The long workdays, kid's school routines, extracurricular activities, and household chores consume each day. Weekends are somewhat of an escape from the weekday grind with occasional downtime but are quickly filled with an endless list of chores and responsibilities.

No wonder the dream of gardening seems out of reach in your hectic life. I hear you, my busy friend—and *I understand the struggle*. However, the reality is you *can* have a garden and still have a life—you just have to be smart and strategic about how you do it!

I'm CaliKim, an organic gardener from Southern California, a working mom, and co-owner of our family-owned business, CaliKim Garden and Home DIY. My husband, Jerry (aka CameraGuy), and I work together to produce content for our YouTube channel and social media that helps people all over the world grow their own organic garden grocery store simply and inexpensively. We also manage a thriving online seed and garden shop that ships garden products globally. Yes, I know what it means to be busy!

Time is a precious commodity; life shows no sign of slowing down and I've had to adapt. Balancing my passion for growing food and tending to my garden alongside family life requires me to work smarter, not harder. This book is a compilation of strategies and tactics I've learned over the years that make the most of my limited garden time.

Regardless of your fast-paced schedule, you *can* have a garden and a life, and I'll teach you the secrets. By making clever choices and maximizing workflow, you can turn your garden dreams into reality. This book is packed with down-to-earth strategies and efficient tactics to optimize your time spent in the garden.

You'll discover how to create your own personal autopilot system, maximizing efficiency and minimizing effort. You'll learn time-saving techniques and strategies that allow you to grow a wider variety of crops, care for your plants effectively, and harvest more veggies in less time than you ever thought possible. As we journey through these pages together, you'll realize that carving out time for a garden is well within your grasp. Yes, you do indeed have time to grow a garden, it's all about being savvy in your approach.

How Saving Time Helps You Be a Better Gardener

Time-saving strategies are not only convenient and help you accomplish more in less time, but they are also crucial for an enjoyable and balanced garden experience.

GARDEN/LIFE BALANCE
Gardening is a fulfilling hobby, and can save you money on groceries, but it shouldn't consume all your free time. Incorporating time-saving tactics into your daily routine allows you to strike a balance between growing your own food and managing other responsibilities, commitments, and social activities.

MAXIMIZED HARVESTS
Timing is crucial for successful harvests. Time-saving strategies like succession planting and organized garden layouts enable you to make the most out of your garden, ensuring a continuous supply of fresh vegetables that saves money at the grocery store.

CONSISTENT CARE = HAPPY PLANTS
Plants thrive with consistent care. Strategies such as automating watering and effective pest/disease control methods ensure your plants have the attention they need to be happy and productive even during busy times.

REDUCED BURNOUT
A well-maintained garden requires ongoing care. Time-saving methods make garden tasks more manageable, helping you garden over the long term without getting burned out.

ENJOYMENT AND RELAXATION
By minimizing time-consuming chores, you can spend more of your garden time observing plant growth, relaxing, connecting with nature, and enjoying the therapeutic benefits of gardening.

How This Book Is Laid Out

The time-saving tactics in this book are divided into bite-sized doable tasks categorized by the amount of time they'll take to accomplish: 3-minute, 5-minute, 8-minute, and 10-minute tasks. This makes it easy to choose tasks based on the time you have available, making it convenient to fit into your schedule.

Each chapter is also segmented into tasks corresponding to the garden's season. This way, regardless of time constraints or the time of year, you'll always have a ready-made list of doable tasks to accomplish on your garden journey.

The Seasons of a Garden

DORMANT SEASON

This season often coincides with winter in colder regions and is when the garden slows down or stops growing due to freezing temperatures. Gardeners with mild winters may not have a dormant season, or if they do, it will be much shorter than in cold winter regions. The garden is often barren and quiet, and it's time for maintenance tasks, planning and organization, and even growing a simple indoor garden of mini veggies, greens, and herbs.

EARLY SEASON

The initial phase of the growing season is generally in early spring. It's the ideal time to prepare the garden for planting, build raised beds, install trellises, amend garden beds, and start planting cool weather vegetables outside.

MID-SEASON

This is the exciting phase when the garden is in full swing and bursting with life. Crops are flourishing and you're savoring daily harvests of fresh produce. This is the busiest time in the garden, and the focus is on harvesting, watering, monitoring for pests and disease, weeding, and general maintenance to keep your garden healthy and happy.

LATE SEASON

The final chapter of the growing year is when summer crops are winding down. Most gardeners are harvesting the last of the tomatoes, cucumbers, peppers, eggplant, and other warm weather veggies and are starting fall cool season crops. In frost-free areas, you can start late season warm weather veggies for another round of harvest before cold weather sets in.

Gardening is a journey, not a destination. Your time is precious. A thriving garden doesn't have to consume all of it. *The 10-Minute Gardener* is your road map to achieving a balance between your garden and other aspects of your life, offering simple techniques and practical time-saving tips to help you grow more food in less time, while still enjoying the life you love.

Time's a tickin', my friend—let's jump in!

CHAPTER ONE

Short and Sweet in 3 Minutes

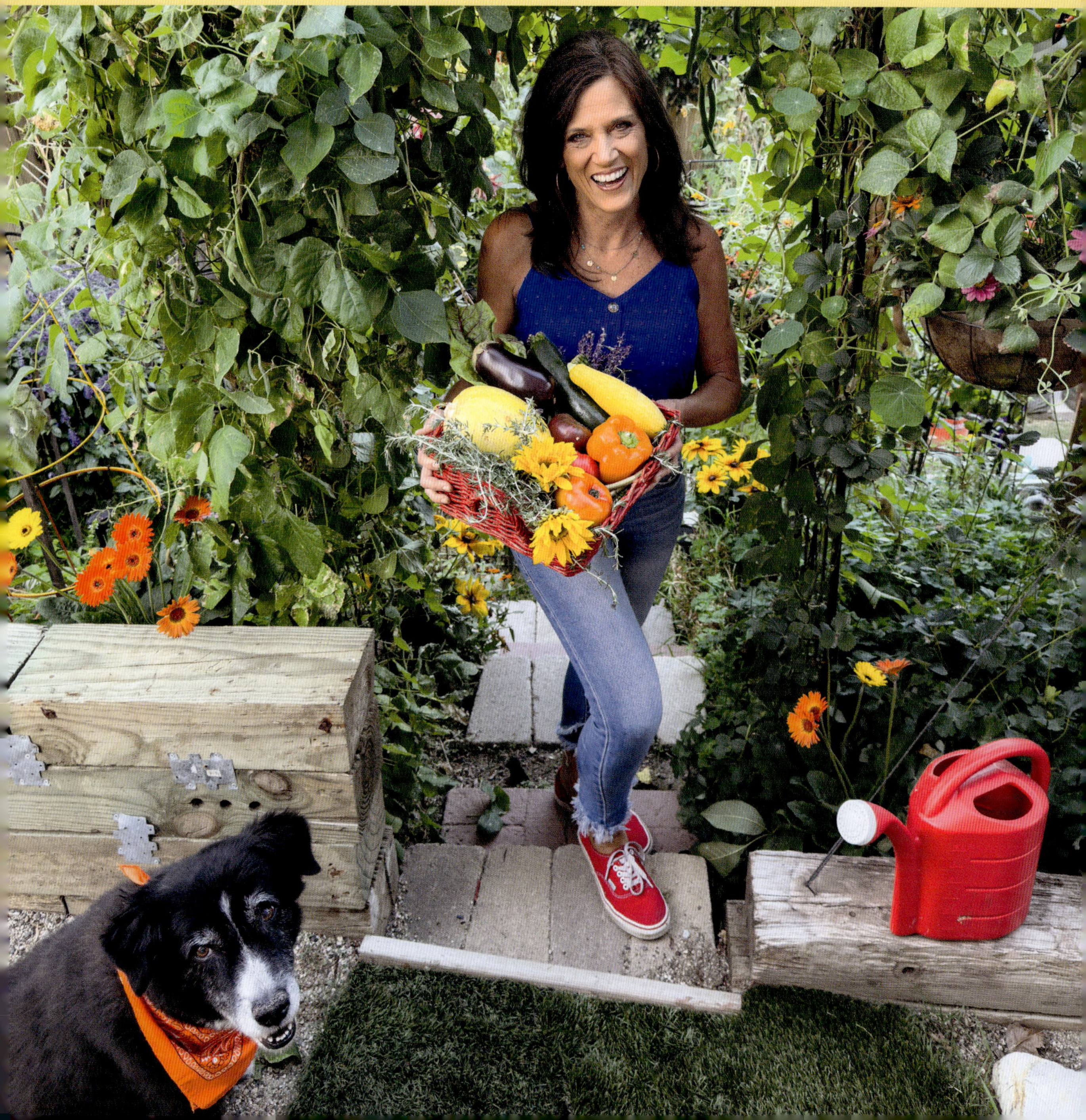

Small Effort—Big Impact

In the hustle and bustle of daily life, carving out time for your vegetable garden might seem challenging. However, no matter how hectic your schedule is, devoting just 3 short minutes a day to your garden can yield remarkable results. These brief intervals, easily incorporated into short breaks or moments of free time, prove invaluable maintaining a thriving garden amid a busy schedule.

Dedicating just 3 minutes daily adds up to 21 minutes per week, resulting in surprisingly significant accomplishments without overwhelming your daily schedule. The key lies in strategic planning—breaking down tasks into short, sweet, manageable moments. You can turn short tasks into golden opportunities to integrate quick garden chores effortlessly into your day, accomplishing big tasks without even realizing it.

Integrating 3-minute tasks into your daily routine makes sure your garden is consistently maintained, even on busy days. Checking on your garden daily for a brief 3 minutes not only helps you catch potential issues before they escalate into major problems, but also contributes to the ongoing beauty of your garden. These quick tasks, ideal for light upkeep and regular check-ins, fit seamlessly into your schedule during moments of free time or short breaks. More complex tasks involving multiple steps are best saved for longer intervals of 8 or 10 minutes.

To maximize your time, keep the list of tasks from this chapter readily accessible for quick bite-sized nuggets of time. By week's end, you'll find satisfaction in checking these short and sweet tasks off your garden to-do list.

Here are some ideas on when to fit these tasks into your day.

MORNING ROUTINE

Use the time when your morning coffee is brewing to water or fertilize a plant, or to deadhead a few flowers, or complete another quick green-thumb task.

DESK STRETCH

If you work at home, take a quick 3-minute break from work to stretch and check off one short task on your garden to-do list.

TRASH CAN TRIPS

Make the most of your trip to the outdoor trash can—every step counts! Grab a quick tomato for lunch, snip some herbs for dinner, water a container plant. A 3-minute delay won't be noticed in the grand scheme of things.

BEFORE DINNER

While waiting for family members to gather at the dinner table, use the time to place a few yellow sticky traps in your indoor garden, propagate an herb cutting, pick a few strawberries, or top off mulch in a container.

BEFORE BED

Wind down your day and clear your mind with a quick note in your garden jotting journal. This screen-free activity will lead to a better night's sleep.

The keys to success are repetition and consistency—integrating these bite-sized tasks into your daily routine and making them automatic habits. This not only contributes to a thriving garden but also allows you to lead a well-balanced life—a true win-win scenario!

Speedy Indoor Pest Prevention Party

In the early stages of a plant's growth, a watchful eye and a few proactive steps work wonders to minimize pesky indoor fungus gnats, spider mites, and aphids. It's far simpler to prevent pests than to eliminate a full-blown infestation. Dedicate 3 minutes two or three times a week to a Pest Prevention Party to ensure your garden is healthy, happy, and relatively pest free.

STEP 1 **Inspect Your Seedlings**
The party kicks off with a quick inspection of your seedlings. Pests love to crash the seedling party, so early detection is key. Take a quick moment to check for any signs of unwelcome visitors: fungus gnats flitting around; webbing, speckled leaves, or tiny red or black dots on leaves (spider mites); or tiny white or green crawling insects (aphids). Be sure to check the underside of leaves too.

STEP 2 **Yellow Sticky Traps**
Place several bright yellow sticky traps strategically around your indoor garden, especially in seedling containers. Fungus gnats can't resist the bright yellow color and get stuck on the sticky surface before they become a nuisance.

STEP 3 **Apple Cider Vinegar Traps**
Create mini-trap stations by filling bottle caps with apple cider vinegar and a few drops of dish soap. Place several around your indoor garden seedlings. Pests are attracted to the sweet scent of vinegar, making the traps an irresistible pest pit stop.

STEP 4 **Keep a Clean Seedling Area**
A spotless, organized seedling area leaves no room for pests to hide. Regularly clear away dry, fallen leaves, plant debris, and spilled soil, and rinse drip trays to eliminate any cozy spots insects might want to set up camp.

STEP 5 **Pest Party Crashers**
For those unwelcome pest party crashers, isolate the pest-infested plant and give it a gentle spray from your kitchen faucet to wash away the bugs. Persistent party crashers? Bring out the big guns—spray with an indoor-safe organic pesticide. Always test spray on a few leaves first to make sure it's safe for your seedlings before you spray the entire plant.

 ## How it saves you time down the line

Investing a brief 3 minutes in a Pest Prevention Party a few times a week translates into substantial time savings later. Early pest detection and preventative measures keep seedlings thriving, minimizing the need for stressful, time-consuming emergency interventions later on. Healthy seedlings grow faster and require less maintenance, and will result in stronger, healthy plants that will be putting homegrown food on your table sooner than anticipated.

Weeding Extravaganza

Transform the often-overwhelming chore of weeding into a lively competition with your partner, friends, or family. Make it a fun-filled event by putting on some upbeat music and see who can gather the biggest basket of weeds to the beat of your favorite tunes. Grab your gloves and head outdoors in the crisp, fresh, early spring air and dig in!

STEP 1 Set a Timer

Setting a timer for 3 minutes turns this daunting task into a bite-sized manageable fun activity. This makes weeding feel doable, especially after a rainy day when the soil is soft and weeds are easier to pull.

STEP 2 Target Early-Season Weed Hotspots

Identify areas where early-season weeds typically pop up and become competitors for water and nutrients essential for young seedlings' growth. Assign each participant in the Weed Extravaganza a container or raised bed to focus on for their weeding.

STEP 3 Quick and Effective Weed Removal

Demonstrate hands-on the technique of pulling gently but firmly upward at the base of the weeds to remove the entire plant, root and all, to minimize the chance of them growing back. You may need to guide family members, especially small children, on the first few Weed Extravaganzas.

STEP 4 Prize for the Winner

After the timer rings, whoever has the biggest bounty of weeds wins! Keep fun prizes like stickers, bookmarks, or healthy snacks on hand. This builds anticipation for the next 3-minute Weed Extravaganza by creating an exciting routine that the family will enjoy.

STEP 5 Weed Disposal

Conclude the Weed Extravaganza by shaking soil off the collected weeds. Say goodbye by tossing them into the green bin. For those with a hot compost pile (it will kill viable weed seeds, toss them into the compost pile to turn them into nutrient-rich fertilizer for your garden!

 How it saves you time down the line

By tackling weeding early in short, enjoyable bursts, you avoid the overwhelming challenge of dealing with giant, established weeds. This sets your seedlings up for success and prevents competition for nutrients and water. The result is thriving vegetable plants, a beautifully maintained garden, and a space for shared family memories and a fun family tradition.

Wise Watering—Efficient Seedling Care

Effective watering is the foundation to seedling growth, and adopting a Wise Watering routine not only helps you remember to water but also helps you learn an important skill for successful early-season seed-starting. This short routine offers step-by-step guidance on how, when, and how often to water your seedlings. With a bit of practice, you'll be watering like a pro, and each session will become more efficient, completed in 3 minutes or less, on autopilot before you know it!

COMMON WATERING MISTAKES

We've all been there—making common errors that hinder our plant growth:

Overwatering: Drowning plants with too much "love" can lead to root rot and mold.

Underwatering: Forgetting to water seedlings can result in stunted growth or plant death.

WATERING FREQUENCY

While there's no magic formula for how often to water your plants, understanding key variables will streamline the task:

Temperature: The growing location's temper- ature significantly affects soil drying time. Warmer loca- tions need more frequent watering, while cooler spots need less.

Planting Medium: Different growing mediums have varying water retention properties. For example, a lightweight water-retentive seed-starting mix will need less watering than a coarser potting mix.

Growing Containers: Smaller containers dry out faster and need more frequent watering than larger ones.

WHEN TO WATER

Water only when your plants need it—they'll tell you when they're thirsty! Look for these two clues:

Color Clue: Dark brown soil indicates moisture; light brown signals drying soil. Allow the top layer to dry a day or so to prevent issues like fungus gnats, then water.

Weight Clue: A heavier container indicates mois- ture; a lighter one suggests a thirsty plant. Lift your seedling container during a 3-minute watering session to assess moisture levels.

HOW TO WATER

Mist: For pre-germination, mist seeds gently with a spray bottle to avoid displacement; once seeds germinate, transition to bottom-watering.

Bottom-Watering: A healthy method for providing even moisture, promoting stronger roots, and preventing disease. It evenly saturates the soil and reduces the frequency of watering.

HOW TO BOTTOM-WATER

▶ Make sure seed-starting containers have drainage holes.

▶ After germination, use a spouted watering can to avoid splashing.

▶ Fill each seedling tray by pouring water into the bottom until it is one-fourth to one-half full.

▶ Let seedlings soak in drip trays of water for 10–15 minutes until the top of the soil turns dark brown, signaling moisture absorption.

▶ Pour off excess water, as standing water causes root rot and attracts fungus gnats and diseases.

CUSTOMIZING YOUR ROUTINE

Each plant has different needs, so tailor your Wise Watering routine accordingly. Observe your plants closely, noting how they respond to different watering frequencies and adjust as needed. Customization makes sure they receive the personalized care needed to become healthy, thriving seedlings.

How it saves you time down the line

Mastering a Wise Watering routine not only means consistent care for your seedlings, but also provides regular check-ins for your plants. This proactive approach minimizes issues like stunted growth and root rot, resulting in healthy seedlings. With no need for seed restarts, seed starting becomes more efficient and significantly less time-consuming.

03
Minutes

Fertilizer Fridays

Welcome to Fertilizer Fridays, where the potentially tricky task of feeding your seedlings is streamlined into a weekly ritual. Mastering this stress-free, simple step-by-step routine is the secret ingredient for a thriving seed-starting process, laying the foundation for flourishing seedlings with consistent growth and robust roots, making your gardening journey more enjoyable and efficient.

EASY DOES IT

To avoid the pitfall of over- or underfeeding, and preventing stunted or yellowing leaves, the mantra is simple: Easy does it! Opt for a mild liquid fertilizer, applying it at one-fourth to one-half strength of the recommended dose. This provides slow and steady growth and avoids burning tender young seedlings.

WHEN TO START

Knowing when to start feeding seedlings is an important aspect of successful gardening. During the initial growth phase, seedlings only grow baby leaves, known as cotyledons. At this stage, they have the nutrients they need stored in their seeds. However, as they progress and reach about 2 inches (5 cm) in height, they reach a developmental milestone. At this point, they develop a set of true leaves, which resemble the final leaves of the vegetable they'll become. This is the perfect time for a nutrient boost that is important for continued plant growth.

WHAT TO USE

When choosing the ideal fertilizer for your plants, opt for a water-soluble, organic fertilizer that dissolves easily in water and provides good root growth, such as VermisTerra Vitality. This earthworm-casting tea stands out not just for promoting strong root growth, but also for added benefits it provides to young seedlings. In addition to a natural growth hormone, it also supplies beneficial bacteria to the soil that fends off fungal diseases. It's also a gentle fertilizer that never burns your delicate seedlings, as many others do. Mix one-fourth to one-half strength with water and feed your plants from the bottom during one of your watering sessions for consistent, healthy growth.

EARLY SEASON

THE 10-MINUTE GARDENER

 ## How it saves you time down the line

Fertilizing is a critical skill that can lead to seedling demise if not done properly. Establishing this weekly routine helps you create a successful seed-starting process, laying the foundation for thriving seedlings with consistent growth. It's an investment in the future health of your plants, making your gardening journey more efficient and enjoyable.

Keep a Garden Jotting Journal

Keeping a garden jotting journal has a multitude of benefits, levels up your garden game, and documents the joys and challenges of your gardening efforts. With a quick 3-minute routine a few days a week, it provides therapeutic and practical benefits.

GARDEN PLANNING

A garden journal serves as your invaluable garden planner, tracking plant varieties, planting dates, watering, fertilizing, and pest control. It becomes your guide for future planting seasons, detailing what worked and what didn't so you can make adjustments for the next season.

DOCUMENTING GROWTH AND ADAPTING

Recording germination dates, growth rate, temperature, light, and other growing condition gives you insights into the unique needs of different crops. For instance, you might note that a tomato took 10 days to germinate at 60°F (15°C), but when the house warmed up to 75°F (24°C) later in the early season, the tomato seed took only 5 days to germinate. This helps you make the adaptation of conditions needed for successful seedling growth.

LEARNING TOOL

An ever-changing learning tool, a garden journal helps you become a more effective, efficient gardener. By tracking different planting times, techniques, and experiments, you're able to evaluate the effectiveness of each, learn from what did and didn't work, and refine your skills with each garden season.

GARDEN THERAPY

Jotting down your thoughts and experiences about gardening is grounding and enriching. It helps you recognize how gardening brings relaxation, stress relief, and joy in connecting with nature. It documents the satisfaction of growing your own food, and the ups and downs—like the excitement of seeing a seedling germinate or the frustration of a plant with problems. It helps you remember the successes and challenges, so you can take pride in your garden journey.

HOW TO KEEP A GARDEN JOTTING JOURNAL
Frequency

Set a routine to jot down garden entries a few times a week. Dedicate a fresh page for each day, dating each entry.

Headings

Create three headings for each entry:

- ▶ What I did in the garden today.
- ▶ What worked, what didn't work.
- ▶ What I learned.

WRITING PROCESS

Under each heading, write a short sentence summarizing your thoughts. Allocate 1 minute for each heading. Don't worry about neatness or complete sentences, as this is for your eyes only.

STREAMLINE

To speed things up, use a voice-to-text memo on your phone. This can be a quicker way to capture your thoughts. The goal is to make the jotting journal quick, easy, and tailored to your personal garden journey.

Monday

1. What I did in the garden today.
   - made a bamboo trellis
   - tomato TLC, picked nasturtiums
   - dash 'n dine harvest

2. What worked/didn't work.
   - worked: harvested tomatoes!
   - didn't work: need thicker twine
     for trellis

3. What I learned.
   - how to grill veggies - yum!
   - prune tomatoes more often

 ## How it saves you time down the line

Journaling your garden experiences is an investment in future garden success. Soon you'll have a record of your early-season activities, providing valuable insights for the next growing season, and a way to track successes and challenges. This ultimately saves time, money, and helps you become a more skilled gardener and enjoy your garden journey to the fullest.

Herb Dash-and-Snip

03 Minutes

Having fresh herbs in the kitchen is one of the greatest joys of gardening. Homegrown herbs add a quick burst of flavor to any culinary creation, whether using mint for refreshing water or tea, basil for grilled veggies, thyme for savory chicken or fish, or chives for flavorful omelets.

As mid-season approaches and your herbs are ready to harvest, integrate a few 3-minute herb dash-and-snip sessions into your weekly schedule.

ENCOURAGES CONTINUOUS GROWTH

Consistent herb snipping stimulates new growth, giving you a steady supply of herbs throughout the growing season. The more you harvest, the more they produce, leading to a steady supply of fresh flavors in the kitchen without waiting for major harvest days.

PREVENTS OVERGROWTH

Snipping off stem tips redirects the plant's energy into producing bushy growth rather than lanky, overgrown stems. This minimizes the need for extensive pruning, keeping your herb garden tidy and consistently producing healthy, flavorful leaves.

PREVENTS FLOWERING

Once herbs flower, their energy shifts from leaf production to seed production, changing the flavor of the leaves. Continuous harvesting discourages herbs from flowering and keeps the plant's energy focused on leaf production, allowing you to enjoy flavorful herb leaves longer.

STREAMLINES MEAL PREPARATION

A constant supply of fresh herbs from your personal garden grocery store streamlines meal preparation. No more last-minute trips to buy packaged herbs, saving both time and money.

PEAK FLAVOR AND AROMA

Freshly harvested herbs mean you'll use them at their peak flavor and aroma, enhancing the flavor of your meals.

REDUCES WASTE

Frequent herb harvesting encourages regular use, minimizing waste. You'll have them at your fingertips in the kitchen and be more likely to incorporate them in your meals.

THE 3-MINUTE HERB DASH-AND-SNIP TECHNIQUE

STEP 1 **Gather Your Tools**

Keep sharp kitchen scissors or precision snips readily available near your herb garden (stick a pair in the soil, pointy side down) to make your quick dash-and-snip sessions a breeze.

STEP 2 **Choose the Right Time**

Set aside 3 minutes a few days a week, preferably in the morning when the oils are the most concentrated for more intense flavors.

STEP 3 **Quick Snipping Technique**

Hold the stem in one hand, scissors/snips with the other, and snip just above a set of leaves or the branching point to encourage bushier growth. Harvest a variety of herbs you'll use in the next 2–3 days.

STEP 4 **Store to Minimize Waste**

▶ For immediate use, pop herbs in jars of water on the kitchen windowsill or counter for quick access while cooking.

▶ For later use, wrap herbs in a kitchen towel or paper towels and place in the fridge, to absorb moisture and prevent spoilage, or wrap stems with string and hang upside down to dry.

 ## How it saves you time down the line

Regular herb harvesting maintains a neat appearance in your garden, avoiding the hassle of dealing with overgrown plants later. Your herbs stay healthier, and you enjoy a continuous supply of fresh herbs in the kitchen to flavor your garden-to-table meals.

Speedy Container Soak

Use the Speedy Container Soak method to transform watering your container garden from a time-consuming task into a time-efficient technique. This is a game-changing method that not only boosts efficiency but also ensures the optimal health and productivity of your container veggies, herbs, and flowers.

Even with drip irrigation in place (see page 180), containers, especially during a heat wave, tend to dry out quickly. Consistent moisture is the key to a thriving container garden, and preventing dry soil that can lead to stunted, lifeless plants is crucial to thriving, beautiful containers. Once the soil dries out, rehydrating with a hose is challenging, as the water doesn't absorb into the soil and simply runs out the bottom.

The solution? Immerse your containers in a storage tote rather than watering with a hose. The soil absorbs moisture from the bottom, becoming thoroughly hydrated, significantly reducing the time spent watering your containers.

The beauty of this technique is that after the quick 3-minute setup, while the containers are soaking, you can take advantage of the time to multi-task and tackle other items on your garden to-do list, maximizing your overall efficiency.

SOAK AND WATER SETUP

▶ This technique is most effective with small to medium containers, 1–10 gallons (4–38 liters). Make sure all containers have drainage holes in the bottom so they can absorb the water.

▶ Use a large storage tote with a flat bottom that is deep enough to accommodate your containers.

▶ Place the containers in the tote and evenly space them for even water absorption.

SLOW AND STEADY SOAKING

▶ Fill the tote with several inches (5–7 cm) of water, just below the rim.

▶ Leave the containers in the tote for 15 minutes to a half hour, allowing the soil time to absorb the water gradually from the bottom for even and complete hydration.

▶ Utilize the soaking time to complete a bonus task, or a few 8-minute tasks.

ADD WATER AS NEEDED

▶ Monitor water level as soil absorbs moisture.

▶ Add more water as needed until the top of the soil turns from light brown to dark brown, indicating soil hydration.

REMOVE CONTAINERS

▶ Once hydrated, remove the containers from the totes and return them to their designated spot in the garden.

▶ The containers will be heavier, signaling sufficient soil moisture.

▶ The soil will stay hydrated longer, reducing watering frequency.

▶ Repeat as needed. An indicator is when the top of the soil turns a light brown color and the top few inches (5–7 cm) of the soil are dry, but before the containers completely dry out.

 ## How it saves you time down the line

Implementing this technique makes container garden watering quick and efficient, reducing the need for constant watering and the time-consuming task of re-hydrating completely dried out soil. Your container plants will thrive with consistent moisture, resulting in more yields of fresh, tasty veggies.

Deadheading Blitz

Mastering the art of deadheading flowers keeps your garden in perpetual bloom and creates a welcoming habitat for our pollinator friends that will also benefit your fruits and veggies! In just 3 minutes, you can tidy up and beautify your garden to make a place for the bees, butterflies, and other beneficial insects to flock to for a pollinator party.

TIME-SAVING BENEFITS
Continuous Blooms
Removing flowers directs energy into producing new buds instead of into seed production, resulting in a longer-lasting display of colorful blooms throughout the growing season.

Neat appearance
Regular deadheading eliminates fading flowers and gives you a vibrant, colorful, well-groomed garden.

Attracts Pollinators
Fresh, colorful blooms attract bees, butterflies, and other pollinating insects that in turn pollinate your garden grocery store.

Visual Delight
You'll delight in immersing yourself in the soothing colors of the garden, seeing the beauty of the pollinators buzzing and flitting about in the garden, an inspiring and motivating sight while deadheading.

DEADHEADING TECHNIQUE
Gather Your Tools
Keep sharp precision snips readily available, strategically placed, pointy side down into the soil in various garden beds for making deadheading quick and simple.

Choose the Right Time
Break deadheading into manageable, frequent sessions, focusing on one garden bed or two to three containers at a time. Set a timer for 3 minutes.

DEADHEADING STEPS
Hold the flower stem in one hand, snips with the other, follow the spent flower stem down the next set of leaves and snip. This avoids "headless" flower stems, promoting bushy growth and more blooms.

How it saves you time down the line

Quick regular flower deadheading maintains well-groomed, colorful blooms, and a welcome haven for pollinators. It means less time spent on major pruning and creates a garden that's both stunning and productive.

Mulch Top-Off

03 Minutes

Mulching your garden doesn't have to be a time-consuming task! Spend just 3 minutes a few times a week topping off the mulch on your garden beds and containers and you'll reap lasting rewards in the garden. Establishing this time-efficient routine in the mid-season puts mulching on autopilot, helps the garden survive summer heat with flying colors, and locks in the benefits of mulch throughout the growing season.

BENEFITS OF MULCH

Conserves Water: A well-mulched garden reduces water evaporation and helps your garden survive dry spells with less frequent watering, saving you both time and water—a win-win.

Root Protection: Mulch shields roots from extreme heat and cold and minimizes plant stress, leading to healthy root growth and increased resistance to disease and pests. More roots mean more fruit!

Soil Health: Mulch gradually decomposes, contributing to soil health by increasing aeration and fertility over time. It also protects soil from erosion, which translates into time and money saved topping off your garden beds with soil each planting season.

Attracts Worms: By keeping soil moist, mulch attracts worms that burrow and aerate the soil, eating the mulch and enriching the soil with natural worm castings.

Weed Suppression: Mulch suppresses weeds by limiting sunlight access to weed seeds, reducing the need for time-consuming weeding.

3-MINUTE MULCH TOP-OFF—QUICK STEPS

STEP 1 Preparation Is Key
Keep various types of shredded mulch in a Compost Sak or trash can for easy access. Keep a 5-gallon (19-liter) bucket near the mulch container to transfer a 3-minute amount of mulch to your garden beds.

STEP 2 Strategic Placement
Position the mulch container near your garden beds or containers for quick and easy application. Keep a 5-gallon (19-liter) bucket nearby.

STEP 3 Set a Timer
Allocate 3 minutes per session, focusing on one to two garden beds or containers at a time.

STEP 4 Mulch Application
Sprinkle mulch 3–6 inches (7.5–15 cm) deep, topping off existing mulch and leaving a small gap around the stems of plants. The more extreme the expected temperatures, the deeper the mulch application should be.

STEP 5 Rake and Even Out
Lightly rake the mulch, spreading evenly, which increases its effectiveness and gives your garden a polished look.

STEP 6 Water
Lightly water to lock in moisture, weigh down mulch, and secure it in place in wind and rain.

 ## How it saves you time down the line

Efficient mulching in quick 3-minute sessions minimizes stress, increases plant resilience, and reduces the need for frequent watering. Less time spent on stressed plants means time saved in disease and pest control, allowing you to efficiently maintain a thriving, sustainable, and eco-friendly garden.

Swift Seed Sowing

Sowing seeds directly in garden beds or containers every few weeks during the mid-season allows you to be proactive in growing a steady supply of fresh, tasty veggies. This strategy spreads the harvests out so you're not overwhelmed with large harvests that might require you more time for preservation than you have available.

This practice, known as succession planting, reduces downtime in the garden and also minimizes waste by growing smaller, more manageable crops.

Quick seed-sowing sessions not only give you a steady supply of produce, but also provide a supply of healthy plants. In the event of pest and disease attacks, you're prepared with newly planted seeds and healthy back-up plants on the way.

IDENTIFY CROPS FOR SWIFT SEED SOWING

Select easy-to-grow, quick warm-season crops your family eats frequently. Keep a running list in your garden journal of veggies, herbs, and flowers, such as cucumbers, beans, leafy greens, squash, basil, sunflowers, and zinnias, that thrive when directly sown in the garden.

KEEP SEEDS HANDY

Keep a basket of these seeds conveniently by your back door. Develop a habit of grabbing a pack of a different variety and putting it in your pocket each time you head out to the garden.

WANDER AND PLANT

During a quick watering session or a leisurely garden coffee walk, look for empty spots in the garden beds or containers. Quickly pop the seeds in the empty spots and cover with soil.

WATERING

No need to worry about watering immediately. Save precious time and let the rain, drip irrigation, or your next hand-watering session handle the initial moisture needs.

THE PAYOFF

By breaking down seed sowing in small, manageable bite-sized chunks, you naturally practice succession planting without even realizing it. As the seeds germinate in the warm summer sun, you'll nurture a new crop of veggies, making sure you have a continual and abundant harvest throughout the growing season. This habit not only saves time, but gives you a consistent satisfaction of seeing the full cycle of gardening from seed to harvest.

⏱ How it saves you time down the line

By sowing seeds regularly, you have a steady supply of fresh veggies, making gardening manageable and efficient. This approach minimizes waste keeps your garden productive, and saves you time, allowing you to enjoy continuous, hassle-free harvests.

Dash-and-Dine Harvest

03 Minutes

Welcome to the Dash and Dine Harvest, a daily late-season forage into your garden grocery store where you'll pick tasty, garden-fresh treats every day. This routine entails quick, daily harvests for a garden-to-table meal—there's nothing fresher or more local than your own backyard, especially when you grew it yourself!

Bring the kids along to experience the joy of picking their own food, and involve them in meal planning to encourage their veggie intake!

ADVANTAGES OF SMALL, FRESH, DAILY HARVESTS

Optimum Flavor and Nutrition: Freshly picked fruits and veggies at the peak of ripeness taste better and have more essential vitamins and minerals.

Greater Variety: Enjoy a diverse range of crops, inspiring creativity with different recipes based on what's available in the garden.

Reduces Waste: Harvesting only what you need for a meal minimizes waste due to spoilage and keeps your produce fresh.

Continuous Production: Regular harvesting encourages plants to grow, leading to consistent production and regular harvests over time.

Gardening Satisfaction: Nurturing a seed from soil to harvest is a fulfilling process that gives you the incredible satisfaction of saying "I grew that!" Sharing homegrown food with those you love is rewarding and satisfying.

HOW TO DASH-AND-DINE HARVEST

Supplies: Keep a water-resistant harvest basket (for quick, in-garden veggie cleanup), pruners, or kitchen shears conveniently near your back door in the garden.

Daily Garden Stroll: Walk through your garden, taking in the vibrant colors, fragrances, sights, and sounds. Look for ready-to-harvest crops.

Harvest: Pick the ripest produce for a well-rounded garden-to-table meal. Fill your harvest basket with a variety of veggies and fruits: a hefty zucchini for grilling; peppers for stuffing; beefsteak tomatoes, crisp leafy greens, and a cucumber for a salad; fresh strawberries for a berry bowl; and a mix of herbs and edible flowers to garnish your culinary creations.

Instant Feast: Head straight to your kitchen to turn your harvest into a delicious, quick meal, a salad, kale smoothie, berry bowl, or grilled veggies—let your garden be your guide to your daily menu.

How it saves you time down the line

The Dash-and-Dine Harvest is a delightful routine that keeps you growing, harvesting, and enjoying fresh meals daily, avoiding large, overwhelming harvests. It keeps meals interesting, lets you tailor them to your family's preferences, motivates and inspires sticking to long-term gardening, saves you time and money on grocery shopping, and provides health and enjoyment for your family.

Sunflower Stalk Trellis

Repurposing a sunflower stalk as a trellis once it's past its prime is a fantastic way to give these tall, beautiful plants a second job to do. It provides a natural, eco-friendly support structure for climbing or vining crops. Creating a trellis from a thick stalk of a dried up mammoth sunflower right where it is growing is a multi-purpose way to use what you already have growing in your garden at the end of the season and makes efficient use of 3 minutes of your time.

MATERIALS YOU'LL NEED

- Old sunflower stalk
- Sturdy pruning shears
- Garden twine or stretchy tie-up tape
- Scissors

STEP 1 Choose the Right Sunflower

Choose a sunflower that has a sturdy stalk (about 1 inch [2.5 cm] in diameter or more) that is starting to dry up and wither. The flowers should have already bloomed and its seeds matured. You can easily identify when the sunflower is at this stage when the flowers are dry and brown, the seeds are visible in the flower head, and they easily fall out when you shake the dry flower.

STEP 2 Cut the Branches off the Stalk

Using pruning shears, cut the branches and leaves off, flush with the main stalk, taking care not to cut the main stalk in the process. When this is complete, the main stalk should be free and clear of any extraneous growth so your climbing plants won't have to compete for space as they grow up the stalk. Leave the main stalk in place in the soil where it is growing.

STEP 3 Secure the Stalk

If the stalk is not securely attached in the soil where it's growing, gently pound it into the ground with a rubber mallet or push it into the soil. It should be anchored firmly about 6 inches (15 cm) into the ground so it's secure enough to support climbing vegetables without toppling over.

STEP 4 Plant Climbing or Vining Crops or Flowers

At the base of the sunflower trellis, plant climbing or vining crops that are lighter in weight, such as peas, pole or runner beans, cucumbers, Malabar spinach, or tomatillos. Pretty vining flowers you can plant include morning glories, trumpet vine, star jasmine, climbing roses, honeysuckle, or sweet pea.

STEP 5 Maintain and Trim

Keep a watchful eye on your climbing plants to make sure they're securely attached to the sunflower trellis. Some plants have tendrils that will naturally grab onto the stalk, while others will need a little help to head in the right direction. Guide small seedlings as they grow toward the stalk and use twine or stretchy tie tape to secure them to the sunflower trellis. Periodically trim any excess growth.

 ## How it saves you time down the line

A sunflower stalk trellis not only provides a functional support for climbing crops and flowers, but is also a natural, visually appealing feature in your garden, making maintenance and harvesting easier. Plus, you'll have the satisfaction knowing that you repurposed a natural resource, saving you time shopping for a trellis at the garden center and saving you money at the same time.

Express Seed Saving

As the garden winds down and your kitchen is overflowing with the late-season harvest, it's the perfect time for seed saving. Allowing different varieties of veggies to naturally dry on the plant or on the vine for seed harvesting is a quick, simple, and effective method of seed saving that not only provides free seeds for next year, but is a beautiful, sustainable cycle of the garden that you can also share with friends.

This technique particularly works well with beans, flowers, and peppers that are easy to dry on the plant, and crops that bolt and flower—such as herbs and leafy greens.

STEP 1 Focus on One Crop Weekly

Select a specific crop weekly for seed saving to keep the task quick, focused, and avoid being overwhelmed.

STEP 2 Harvest Seeds at the Right Time

▶ Choose the right time when seed pods or flower heads have dried completely on the vine and reached full maturity.

▶ Check pods and flowers; they should feel and look dry and crispy.

▶ Harvest before pods and flowers split open, ensuring the seeds are at peak quality, preventing moisture from rain.

▶ Use scissors to cut the pod or dry flowers from the plant.

STEP 3 Extract the Seeds

▶ Make sure your hands are clean and dry to prevent getting moisture on the seeds.

▶ When pods are fully dry, remove the seeds from the pods by splitting open to extract the seeds.

▶ Crumble dry flower heads between fingers, letting the seeds fall on a plate or tray.

STEP 4 Separate Seeds from Chaff

Use a sieve to separate seeds from the dry chaff.

STEP 5 Seed Storage

▶ Store dry seeds in a completely airtight glass jar, or a sealed bag.

▶ Label each container with the variety and seed harvest date for accurate seed management.

▶ Store the seeds in a cool, dry place to maintain seed viability.

How it saves you time down the line

Over time, saved seeds from your garden adapt to your specific growing conditions, reducing the need to search for new varieties that will do well in your climate. This results in hardier, disease-resistant crops better suited to your area, saving time experimenting with new varieties. Saving your favorite and most productive varieties allows you to grow them year after year and share seeds with friends. This saves time on seed shopping but also reduces the money spent on seeds each growing season.

Relax and Recharge Break

03 Minutes

In the midst of our bustling schedules, it's easy to forget to savor the fruit of our labors. Incorporating the time-saving tactics in this book into our routine ensures we can have both a flourishing garden and a life. Remember, the garden is a perpetual work in progress—embrace it. Make it a daily priority to relax and recharge—even if it's just for 3 minutes.

Especially during the late season, the garden time is busy with harvesting, pruning, and managing end-of-season pests and disease. Taking a few moments to relax and appreciate the beauty that you've worked hard to create is crucial.

BENEFITS OF A 3-MINUTE RELAX AND RECHARGE BREAK

Stress Relief: Nature's therapy is the best therapy! The garden's green calming vibes melt away the stress and ease tension accumulated during the day.

Mindful Escape: Step away from screens and noise. The garden offers a peaceful retreat, helping you to be present in the moment and connect with nature.

Renewal: Recharge your batteries! Spending time in the garden boosts your mood, inspires and motivates you, avoids garden burnout, and leaves you feeling refreshed for challenges ahead.

DAILY RITUALS FOR QUICK RELAXATION

Morning Coffee Walk: Start your day with a cuppa joe and wander around the garden to start your day. No chores, just pure enjoyment—observe and enjoy your plants and soak up the morning magic.

Mid-Day Music Break: Pause your to-do list, step outside, put on your favorite tunes, and sit for 3 minutes at your favorite "chill" spot in the garden. Observe the beauty of the flowers and the buzzing pollinators in the garden's beauty.

End-of-the-Day Wind Down: Gather the family after dinner for a quick garden wind-down. Whether it's to enjoy the beauty of a golden-hour sunset, an after dark bug hunt, or sitting under a summer evening moonrise, enjoy the shared moments and savor the beauty.

Whichever garden escape you choose, relish the beauty around you and bask in the joys of being a gardener, just as nature intended. This is garden therapy at its best!

How it saves you time down the line

In the rush of busyness, neglecting self-care leads to decreased motivation and burnout. Recharging makes sure you stay eager and motivated for long-term gardening and avoids wasted time procrastinating due to burnout.

3-Minute Tool Tidy

The dormant season in the garden is the perfect time for maintenance tasks that are often overlooked during the busy peak season. At the top of the list is a quick tool tidy. Incorporating a 3-minute weekly tool tidy routine throughout the dormant season means you'll kick off the busy growing season with maximum efficiency. The time investment accumulates, setting the stage for a flourishing spring garden.

Maintaining well-organized and properly functioning tools has a range of benefits:

Reduced Stress: A tidy toolshed or storage area reduces stress of a cluttered environment, increases focused gardening time, and is a visual motivator.

Easy Access: Organized tools translate into quick access, less time searching (yes, we've all done that!), and more time tending to your garden, increasing overall efficiency.

Long-Lasting Tools: Proper storage and regular maintenance prevents tools from rusting or deteriorating, extending their lifespan and saving you replacement costs. Keeping tools tidy is simple with regular attention. Choose one 3-minute task to complete each week:

Quick Clean and Wipe Down: Wipe away season build-up, removing dirt, debris, and rust. Use rubbing alcohol on pruner blades to sterilize, and hose off shovels, rakes, and other large garden tools.

Sharp Blade Check: Inspect and sharpen the cutting edge of your pruners, shears, and other cutting tools for optimal performance.

Oil Moving Parts: Apply a light coat of oil to moving parts of pruners and shears to prevent rust and maintain smooth operation.

Tool Organization: Organize your tool storage area, making sure each tool has a designated place. Here's a quick and easy idea: Set up a 5-gallon (19-liter) bucket with tools you use daily. This ensures efficiency during regular garden maintenance because the tools are accessible and portable to move around the garden with you. At the end of the day, they're conveniently already in their spot, ready for the next use.

 ## How it saves you time down the line

By maintaining and organizing your garden tools, you level up your garden efficiency by saving time searching for tools and eliminate frustrations with nonworking tools, making your garden experience enjoyable and fun.

Strategic Seed Inventory

Empower your off-season garden planning with a simple but powerful seed inventory strategy. It's not just about knowing what seeds you have, it's also a game-changer for time- and cost-efficient garden planning for next season.

Here's how to integrate a simple yet strategic seed inventory routine seamlessly into your garden schedule once or twice a week.

3-MINUTE TIMER

Set your timer for short and sweet 3-minute increments. This short burst maintains focus and prevents you from feeling overwhelmed. Tackle one section of your seed storage area in each session, breaking down the task into manageable bite-sized chunks.

GATHER AND LIST

In those quick 3 minutes, gather leftover seeds from the previous season and make a detailed list of what you already have. After several seed inventory sessions, you'll have a clear understanding of what you already have and what you need for next season.

CONSISTENCY IS KEY

Repeat these sessions one to two times each week. The consistency will gradually build up a comprehensive list of what you have, which paves the way for informed seed purchases.

INFORMED GARDEN PLANNING

Armed with knowledge of your existing seed inventory, your spring garden planning becomes strategic and informed instead of haphazard. No more last-minute scrambles for the right seeds at the right time. You'll eliminate duplicate purchases, increase cost effectiveness, and have the seeds you need to start at the time you need to start them.

STAY THRIFTY

Take advantage of off-season clearance sales at garden centers and online seed shop. These seeds are perfectly fine, they're often just making room for fresh varieties. This thrifty approach gives you more varieties to grow without breaking the bank.

BROWSE SEED CATALOGS

Enjoy the delight of browsing through seed catalogs—eye candy for gardeners! They are not only a visual feast, but also provide inspiration, motivation, and ideas for exciting new varieties to grow in the upcoming season.

 ## How it saves you time down the line

As early season approaches, your comprehensive seed inventory means you'll be ready with the seeds you need for efficient seed starting in the early season. This straightforward routine not only prevents overwhelm, but also allows you to make the most of your garden downtime, and helps you enjoy planning for the next growing season.

Indoor Sprouting— Beat the Winter Blues

The dormant season is the perfect time of year to begin a mini indoor garden to enjoy fresh veggies during the winter. Growing fresh herbs, leafy greens, microgreens, and sprouts indoors is simple, fun, and a winter must for any green thumbs.

NO SOIL, NO GROW LIGHTS—NO PROBLEM!

Growing sprouts indoors is a game changer for indoor gardening. It's a quick and simple process, requiring minimal time, space, and equipment. Adding a short and sweet 3-minute sprouting routine to your daily schedule not only provides a huge a nutritional boost, but also a mood lift on those long winter days.

CHOOSE YOUR SPROUTING SEEDS

Your indoor garden sprouting adventure begins with selecting the right seeds. Opt for easy-to-grow varieties like lentils, legumes, mustard greens, and radishes. These seeds not only sprout quickly, but also offer a wide array of flavors, textures, and nutritional content. Experiment with different seeds to discover your favorite—the possibilities are endless.

STEP 1 Soak Seeds

▶ Place 1–2 tsp (5–10 ml) seeds in clean, clear jar.

▶ Use a larger jar for more sprouts and adjust seed quantity accordingly.

▶ Cover seeds with warm water.

▶ Place a mesh sprouting lid on jar.

▶ Soak seeds 6–8 hours or overnight.

▶ The soaking process breaks seed dormancy and kickstarts germination.

STEP 2 Rinse and Repeat

▶ After initial soaking, turn jar upside down (keeping mesh lid on) to drain excess water.

▶ Position jar upside down at an angle on a sprouting stand or on the kitchen counter (no grow lights needed) to continue draining.

▶ Rinse seeds two to three times each day, immediately draining each time.

▶ Seeds will begin to sprout after 2–3 days.

▶ The routine of rinsing and draining keeps the sprouts moist while preventing bacterial or mold issues.

▶ Continue this routine two to three times a day until harvest.

STEP 3 Harvest and Enjoy!

▶ Within 3–5 days, the seeds will transform into fresh, delicious, and nutrient-packed sprouts.

▶ Harvest when most of the seeds have sprouted and the jar is one-half to three-fourths full.

▶ Remove sprouts from the jar and enjoy right away for maximum freshness and flavor.

▶ What you don't eat immediately, wrap in paper towels or in a sprout storage bag to absorb moisture and extend their shelf life. They are best if consumed within 2–3 days.

▶ Enjoy your sprouts on sandwiches, salads, soups, eggs, chicken, fish—the sky's the limit!

▶ For a continual harvest, start a new jar of sprouts every 3–5 days.

 ## How it saves you time down the line

Maintaining a simple indoor green routine during the dormant season when it's too cold to grow outside helps beat the winter blues and gives a connection to nature that gardeners crave. Even if you can garden year-round, having fresh produce at your fingertips encourages heathier eating habits and provides an enjoyable daily routine. Keeping the garden spirit alive in winter provides inspiration and joy, leading to more productive and enjoyable use of your time.

3-Minute Spring Daydream

As winter and the dormant season settles in, with shorter days, often the winter blues can hit and you're itching to get out in the garden again. Rather than getting down in the dumps, turn the downtime into an opportunity to plan and dream of the next growing season. Dedicate a few 3-minute sessions each week to record your thoughts and garden dreams in your garden jotting journal.

WHAT TO INCLUDE IN YOUR 3-MINUTE SPRING DREAMING

Journal Your Garden, Bed by Garden Bed

Reflect on what worked well and what didn't work during the past growing season. Record these insights in your journal. Let this be your inspiration and guide for your upcoming spring garden.

Make a Produce List

Make a list of the fruits, veggies, herbs, and flowers you want to grow in each garden bed and container. Include tried-and-true family favorites, as well as new varieties you'd like to experiment with.

Sketch Your Garden Layout

Create a simple, rough sketch of your garden space. Remember, it's not meant to be an art masterpiece—this is for your eyes only and will help you visualize your future garden. Decide where you want to plant specific crops and include space for future garden dreams, like interesting garden structures, seating areas, trellises, or new raised beds you'd like to incorporate into your garden.

Plan Your Garden Calendar

Make a calendar to outline your garden plans for each week of warm weather. This is a strategic tool to help you make the most of your garden time. Take into consideration what needs to be planted when and refer to resources such as my succession planting chart in *The First-Time Gardener: Raised Bed Gardening* to help you decide which crops to start from seed indoors, which crops to direct seed, and about when you can start them in your area.

Picture Your Dream Garden

Take screenshot pictures of your dream garden beds or garden elements and choose one or two that are a good fit to incorporate into next season's plans.

 ## How it saves you time down the line

Thinking through garden planning step-by-step during the dormant season without the pressure of the early spring season offers space and time for your mind to wander and sparks creativity. By the time spring arrives, you'll have an entire journal filled with garden ideas and can focus on the ideas that are the most feasible for you, saving you the pressure of coming up with ideas at the beginning of a busy growing season. This maximizes your time and will help your spring garden thrive and be space efficient.

Short and Sweet Container Care

03 Minutes

Take advantage of the time in the dormant season with a few 3-minute sessions to clean up and tidy your containers and seed-starting supplies. With minimal effort, they'll be cleaned up in no time, ready and waiting for next season.

FABRIC PLANTERS

Durable reusable fabric planters (such as Smart Pots) can be tossed into the washing machine and air-dried so they have a fresh clean look at the beginning of the next growing season. Empty out soil into a spare trash can, rinse out residue (upside down over a 5-gallon [19-liter] bucket makes quick work of rinsing), toss into the washer, and air-dry. Store clean and dry containers in a bin in your garage so they're ready to rock and roll when weather warms up.

CERAMIC PLANTERS

If you live in an area with freezing winter temperatures, empty soil out of ceramic containers to avoid them cracking during the winter. Wash with warm soapy water and store in a sheltered location to protect them from the elements.

SEED-STARTING CONTAINERS

Rinse six-pack seed-starting containers and drip trays so they are ready for reuse. Empty out and save soil in a designated location. A quick blast of water from the hose removes soil debris. After rinsing most of the soil, dip them in a soapy tub of warm water to clean, removing any tiny bug eggs or bacteria. Air-dry and store in the garage or toolshed to protect them from the elements. This saves money by reusing valuable resources that are still in good working condition.

 ## How it saves you time down the line

Keeping your containers in good condition avoids taking the time and expense of purchasing new ones each season. In a few short container care sessions, you'll have clean and ready-to-plant containers in a central location at the beginning of the next growing season. Plus, you'll enjoy a tidy and well-organized garden space, avoiding the confusion of cluttered space, clearing your mind, and paving the way for increased garden productivity.

05
Minutes

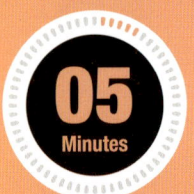

CHAPTER TWO

Mini but Mighty in 5 Minutes

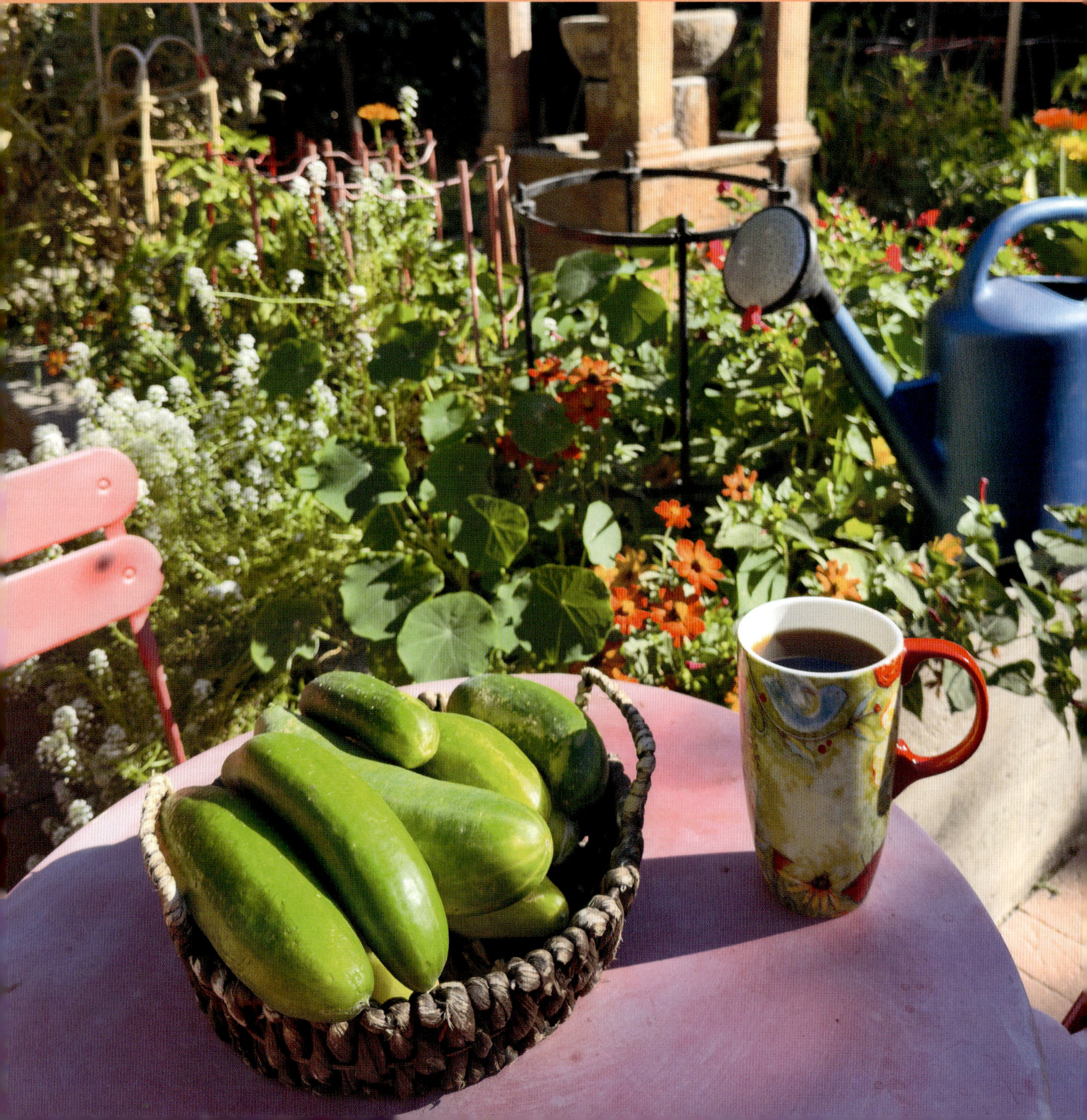

Maximize Your Moments

There's a wealth of potential in just 5 short minutes, especially when used strategically. Unlocking this potential lies in recognizing and utilizing those mini pockets of time wisely, so without even realizing it, you've accomplished a lot.

Although 5 minutes is a small amount of time, mighty results can be accomplished when multiple segments are added up. Completing four 5-minute tasks daily adds up to 20 minutes each day, 140 minutes weekly, or 2 hours and 20 minutes of gardening progress, often without even noticing it!

Squeezing mini 5-minute garden tasks into your day takes surprisingly little effort. Many short activities can be seamlessly integrated into your daily routine, making the most of brief moments during the day, without feeling like chores. Make them part of your daily rhythm, so you're operating on autopilot.

Here's some ideas on how to weave them into your day.

MORNING COFFEE ROUTINE
Nothing like kicking off your day with a dose of garden therapy. Brew yourself a morning cuppa joe and step outside for a few minutes. You'll enjoy the fresh morning air, get a great start to your day, and accomplish a "chore" before you even realize it.

LUNCH BREAK
Use your lunch break to refresh your mind and connect with nature. Water your container plants, grab some berries or cucumbers for lunch, and check a few quick tasks off your garden to-do list.

AFTER WORK WIND DOWN
The garden is calling to you after a long work day—can you hear it? Winding down with light garden tasks such as watering or harvesting helps you de-stress and transition from work mode to home mode.

WHILE ON THE PHONE
Multi-task with garden chores like watering, deadheading flowers, or pruning while taking a phone call.

COOKING DINNER
There's nothing fresher and more local than your own backyard. Dash out for a quick harvest of herbs and greens for a garden-fresh salad for dinner and avoid a stop at the grocery store on the way home from work.

While Waiting for the Kettle to Boil

Steal the 5 minutes waiting for the kettle to boil to check on your indoor plants, water, feed, or prune, do a quick repot, or tidy up your indoor garden.

Tapping into the power in 5 minutes lies in identifying and utilizing those short pockets of time wisely. Keep a list of 5-minute tasks ready to roll. Plan ahead and incorporate the tasks in this chapter into your daily routine to efficiently maintain your garden, even on hectic days. Take pride in nurturing your garden throughout the day, enjoying the satisfaction of accomplishing significant tasks in multiple small but mighty 5-minute increments.

Tie-up Tuesday/Thursday Routine

Are you ready to transform your backyard or patio into a thriving, space-efficient veggie garden? Grow vertically! Utilizing trellises not only maximizes yields and minimizes space, but also enhances air circulation and keeps your plants healthy, happy, and off the ground away from pests and critters. Growing your veggies vertically allows you to easily monitor their progress, and quickly harvest ripe vegetables instead of searching for them under sprawling vines.

ESTABLISH A TIE-UP TUESDAY/THURSDAY ROUTINE

Don't let trellising your plants get out of hand and become a daunting task. Instead, establish a Tie-up Tuesday/Thursday routine. On these days, dedicate just 5 minutes tying up your vining veggies—whether it's before work, walking your furry friend, or while waiting for dinner to cook in the oven. Consistently practicing this quick routine transforms your garden into a thriving, space-efficient paradise that's producing an abundance of veggies during the peak season.

ESSENTIAL TOOLS AND SUPPLIES

Set yourself up for success by gathering essential tools and supplies that make this process seamless. You'll need twine, string, stretchy tie-up tape, scissors or pruners, bamboo poles, wooden stakes, rebar, T-posts, wire, pliers, and a heavy rubber mallet. Having these items at your disposal ensures that you can breeze through your tie-up routine.

STRATEGIC PLACEMENT

Strategically place these items around your garden for maximum efficiency. In my garden, stretchy tie-up tape is hanging unobtrusively in garden beds, and pruners are placed in containers or in hanging baskets along pathways. Shorter bamboo poles and wooden stakes are placed along fences and at the back of garden beds. Heavier supplies are tucked in a shed or accessible corners of the garden.

PRE-CUT

Before your tie-up session, grab your twine and pruners from their strategically placed location. Cut several lengths of twine or stretchy tie-up tape and hang them from your belt loop for easy access.

START EARLY

Start the trellising process early in the season when your climbing crops are seedlings. Gently guide the main stem of young seedlings toward the trellis as they grow, setting the stage for upward growth.

SECURE THE PLANT—BEGIN AT THE BASE

As your plants begin upward growth, start tying up at the base of each plant. Use your pre-cut twine or tie-up tape to make a knot around the stem and attach it to the trellis. Don't tie the stem tightly; keep it loose to allow the stem flexibility and room to grow.

REPEAT AS NEEDED

As the plant grows and develops multiple stems or branches, repeat this process every 4–6 inches (10–15 cm) along the stem, always starting at the base, so the plant is adequately supported from the ground up.

PLANT EMERGENCY FIX FOR OUT-OF-CONTROL PLANTS

We've all been there—plants that missed early care, outgrew your expectations, or simply became unruly. To save them from toppling over, grab a wooden stake from your garden bed's corner, pound it near the main stem, and tie it up. Larger plants may need several stakes or T-posts. Encircle the plant with heavy-duty twine tied to the stakes.

 ## How it saves you time down the line

This quick routine provides plant support in the early season, resulting in a neat and tidy garden instead of a mess of unruly plants. Trellising veggies as they grow is easier than tying up sprawling plants. Vertical growing increases air circulation and reduces the need for disease/pest control spraying in the later season, allowing you more time to savor the beauty of your garden and enjoy quality time with your family and friends.

Express Hardening Off = Happy Seedlings

As the last frost date approaches, you're eagerly anticipating homegrown veggies on your dinner plate. Eager to get indoor seedlings in the garden, you excitedly plant them outside and see them wither and die. Why? You missed a crucial step: hardening off.

WHY IS HARDENING OFF SO IMPORTANT?

Pampered indoor seedlings aren't ready for harsh outdoor elements, and sudden temperature changes can be disastrous. Hardening off is a gradual acclimation process so your seedlings are happy and ready for the great outdoors. Not a lengthy or complicated process, this small but mighty step is the key to outdoor seedling survival. Master this process and your seedlings will thrive in the early season. Let's dive into the 5-minute-a-day express version with time-saving tips—a manageable and stress-free routine!

STEP 1 Check Average Last Frost Date

Use almanac.com or an online frost date calculator to check your last frost date. Start hardening off a week before you'll be planting outside. Cool weather veggies can be planted outside just after your last frost date; warm weather veggies will follow 2–3 weeks after that as the soil temperature warms up.

TIME-SAVING TIP: Set a calendar alert for your upcoming last frost date for proactive scheduling.

STEP 2 Day 1—Select a Shady Spot

Select a shady spot or begin on a cloudy day. Place seedlings outside for 1–2 hours, then bring them back indoors under grow lights.

▶ Choose a spot close to your house for convenience.

▶ Group similar seedlings together for efficiency.

▶ Start with just two to three seedling trays weekly that you can move in and out in a few minutes.

▶ Working? Start on the weekend or a day when you're home early for efficient monitoring.

TIME-SAVING TIP: Use a timer on your phone to avoid constant clock watching.

STEP 3 Day 2—Sunlight Exposure

Expose seedlings to 1–2 hours of sunlight, then back in the shade for 1–2 hours, extending total outdoor time to 3–4 hours, then return indoors under grow lights.

TIME-SAVING TIPS:

▶ Multi-task by planning sessions during other outdoor activities to easily monitor plants for signs of stress.

▶ Use a weather app with notifications on to monitor the forecast. Bring plants indoors if there's a sudden change.

STEP 4 Days 3–7—Gradually Increase Sun Exposure

Gradually increase sunlight/daylight exposure each day until seedlings are outside all day long, then introduce to nighttime exposure. Monitor the weather to avoid chilly or frosty nights. After they have survived a night outdoors with flying colors, they are garden-ready!

TIME-SAVING TIP: Use weather apps with automatic alerts to monitor nighttime temperatures.

⏱ How it saves you time down the line

Dedicating 5 minutes a day for 7–10 days in the early season is a small but mighty time-saving investment. Having resilient, disease-resistant plants minimizes the time-consuming process of re-starting seedlings or nursing struggling plants back to health. Robust, healthy plants mean quicker harvests and putting veggies on your table sooner.

Speedy Container Prep

Container gardening is the quickest way to turn your space into a thriving vegetable garden. Use fabric containers for quick setup—unfold, pop up, add soil, and plant your hardened-off seedings for an instant garden.

Spend 5 minutes a few times a week to have some containers ready when your seedlings finish hardening off.

SUPPLIES

Organization and pre-prep make for time-efficient container filling. Keep these supplies in a convenient location near your container garden space:

▶ 5-gallon (19-liter) fabric containers

▶ DIY container soil or bagged potting mix

▶ Hand tools: shovel, rake

▶ Storage totes

▶ Worm castings, organic fertilizer for reused soil

▶ Water

FILL CONTAINERS

▶ Prep at least two containers simultaneously for efficiency.

▶ Set up an assembly line to add soil and amendments.

▶ Fill containers by folding down the fabric sides and gradually unfolding as you add soil.

PRE-MOISTEN

▶ Pre-moisten soil prior to planting for even water distribution when planting.

▶ Soak soil-filled containers in storage totes filled with water. The soil absorbs moisture through the bottom of the container for thorough hydration.

▶ Soil is moistened when the top is dark brown and the container is heavier, indicating sufficient moisture content.

▶ Remove the container from the storage tote.

▶ When reusing container soil, there is no need to dump out the soil. Loosen the existing container soil, mix in 2–3 inches (5–7.5 cm) of fresh potting mix, a handful of organic granular fertilizer and worm castings, and follow the above steps to hydrate.

CLEAN AS YOU GO

▶ Tidy up as you go to prevent a cluttered work space.

▶ Clean up spilled soil, and return tools and unused soil back to their proper places.

▶ This makes clean-up quick and easy, and gives you a fresh start for the next container prep session.

How it saves you time down the line

Proper container preparation is a time-saving investment. Soaking filled containers in storage totes to pre-moisten is a huge time saver over pre-moistening by mixing water into soil prior to filling. Prepping a few daily means you'll have containers ready when seedlings are ready to plant, preventing overgrown, stunted plants and minimizing time spend re-starting seedlings.

Watering Tune-Up

In the gardening journey, the art of watering is sometimes a trial and error process, but a critical one to master for a garden that produces lots of fresh, delicious veggies to fill your dinner plates. Early-season plants require tender loving care for optimal health. Investing just 5 minutes in a short watering tune-up a few times a week yields significant dividends in terms of steady plant growth and strong root development, even during dry or rainy spells.

GARDEN WATERING STROLL

Set a timer for 5 minutes and take a stroll through your garden, perhaps on a lunch break or with morning coffee in hand. This simple, yet satisfying routine allows you to keep a pulse on your garden's watering needs and make adjustments, whether to increase moisture in dry spots or reduce watering in areas that are oversaturated. Focus on just one or two garden beds or containers during your watering tune-up stroll to prevent overwhelm and quickly finish the task in the allotted time. Look for dry or overwatered soil and keep an eye out for signs of seedling stress, such as wilting or yellowing leaves. Ideally the top few inches (5–7 cm) of the soil should be evenly moist, not muddy or waterlogged.

MOISTURE MANAGEMENT

Water your plants only when they need it. Use your finger as a moisture meter by inserting your finger into the soil. If the soil feels moist, hold off on watering. If it feels dry, it's time to water. This hands-on practice makes sure that your watering routine is guided by the specific needs of your plants.

EASY DOES IT

Give special attention to young seedlings, watering them at the base of the plants with a gentle sprinkle to avoid disturbing their delicate roots. Adjust your drip irrigation timer schedule as needed to ensure that each watering zone provides an appropriate level of moisture to plants.

 ## How it saves you time down the line

This 5-minute watering tune-up routine not only fosters even plant growth, but also prevents stunted development due to under- or overwatering. This peaceful, satisfying routine of tending to your garden becomes not just a task to check off your garden to-do list, but a process of nurturing a thriving garden that will in turn feed and nourish you and your loved ones.

Time-Saving Compost Turning

To accelerate your compost pile's break down, turn it every 3–4 days (see page 174 for how to build a compost pile) to aerate it and keep the microbes happy. If the pile is dry, add water and mix in for even moisture.

Maximize efficiency with these time-saving tips and techniques:

KEEP TOOLS HANDY
Store a pitchfork or compost-turning tool nearby for quick and easy turning. A 5-gallon (19-liter) bucket or wheelbarrow helps with overflow or transport of compost materials.

STRATEGIC LOCATION
Place compost near your house for convenience. You're much more likely to turn your pile weekly if it's near your backdoor than in a corner of your backyard.

SIZE MATTERS
Opt for a smaller pile (like a 50-gallon [190-liter] Smart Pot Compost Sak) for easier turning. Divide larger piles into two smaller piles for quicker and more manageable turning. Turn one pile in each 5-minute session.

INSIDE-OUT TECHNIQUE
Use a pitchfork to shift materials from the outside to the warmer inside for even decomposition. If the pile is too dry, sprinkle with water; if too wet, add dry materials, like shredded leaves or straw.

AERATION TECHNIQUE
Instead of turning the entire pile, poke it with a metal pole or pitchfork to lift and loosen materials. This aerates without turning the entire pile, but isn't as physically demanding.

FREQUENCY IS KEY
Do once-a-week, 5-minute smaller turns instead of monthly turns to prevent compaction. This also allows you to effectively monitor moisture and accelerates decomposition.

How it saves you time down the line

Weekly 5-minute turning has a big payoff in the form of cost-effective black gold for your garden. Regular, smaller turns prevent compaction, maintains optimal conditions for microbes, and keeps the compost process enjoyable and manageable. This saves effort, time, and money and provides nutrient-rich compost to feed your garden sooner.

Tomato TLC in a Flash

Tomatoes, the most beloved of garden veggies, inevitably face disease during the growing season. Disease often creeps in during the mid-season. This doesn't mean you're doing something wrong, it's just part of the tomato journey. However, in just 5 minutes a few times a week, you can master essential tomato pruning techniques to enhance airflow, minimize fungal diseases, and boost harvests. This quick tomato TLC routine keeps your tomatoes manageable, healthy, and flourishing.

TECHNIQUE #1: PRUNE FOR AIR FLOW BOOST

Targeted Pruning: Prune branches on the bottom 6 inches (15 cm) of the plant to minimize the risk of foliar and soil-borne diseases splashing onto the plant.

Efficient Cuts: Keep clean, sharp pruners near the pruning area to maximize cutting effectiveness and minimize time searching for tools. Cut each branch close to the main stem, taking care not to damage it, to boost air circulation and to provide easier access to the base for plant management.

TIME-SAVING TIP: Work smarter, not harder, by keeping a compost bin for disease-free branches and a separate trash bin for diseased branches near your pruning area.

TECHNIQUE #2: PRUNING SUCKERS FOR LARGER, HEALTHIER FRUIT

This technique is most effective on indeterminate tomato varieties that produce all season until frost, not determinate varieties.

Quick ID: Learn to identify suckers quickly by focusing on the shoots that grow out of the "armpit" or the main juncture between the main stem and branches. These suckers divert energy away from fruit production if left unchecked. Pruning them redirects the plant's energy into fruit production and results in larger, heathier fruits and a manageable plant size.

Weekly Consistency: Consistent, brief weekly sessions avoid the overwhelming, time-consuming task of pruning overgrown plants.

TIME-SAVING TIP: Keep pruners sharp, clean, and strategically placed near tomato plants to reduce time spent searching for tools.

TECHNIQUE #3: DISEASE DEFENSE— PRUNING YELLOWING LEAVES

Early Detection: Train yourself to always be on the lookout for yellowing or spotted leaves, often the first sign of tomato disease. Early detection and removal of these leaves/branches can mean the difference between a tomato plant succumbing to disease mid-season or being productive throughout.

Vigilance Is Key: Integrate disease defense into your regular tomato TLC routine. This proactive approach slows down tomato disease without the time commitment of additional pruning time down the line when the disease is more advanced.

 How it saves you time down the line

Consistency in this small but mighty tactic establishes a routine that prevents further issues, resulting in a more manageable garden. It demands less maintenance, time, and effort in the long run, yielding healthier, more productive and delicious tomatoes. Spend more time enjoying the fruits of your labors and less time dealing with the drudgery of diseased tomato plants.

Shade Cloth Pop-Up— A Mid-Season Oasis

As the mid-season summer heat wave hits, even our favorite warm-weather-loving veggies like tomatoes, peppers, cucumbers, and squash can overheat in the scorching sun. When 90°F–95°F (32°C–35°C) temperatures hit, the blossoms dry up and drop off, stalling production. The plants struggle to stay alive and redirect energy into survival rather than fruit production.

A quick and efficient shade cloth pop-up routine is a literal life-saver for your garden, shielding your plants from heat stress and putting food on your table quicker.

PRE-PREP FOR SUCCESS

Prepare for the summer heat by having 40–60% shade cloth (blocking the sun's rays) ready to roll. This pre-planning avoids the last minute scramble and makes sure you're ready for quick and efficient pop-up when needed. Invest in high-quality, durable shade cloth to minimize UV damage and wear and tear from the wind, ultimately saving both time and money in replacement and repairs. This strategic pre-prep time translates into a 5-minute shade cloth pop-up on hot days.

Cut multiple shade cloth pieces to fit different garden areas, and label each piece for even quicker setup.

PORTABLE SOLUTIONS

Opt for portability over permanent structures for attaching shade cloth. Clip it to trellises on raised beds or use stakes in containers for quick and easy setup on hot days. Leave shade cloth clipped to the corners of raised bed trellises so you can quickly attach to the other side for shade cover over the entire bed in minutes. Invest in a weatherproof storage container with easy garden access so you can neatly fold and store shade cloth when not in use to minimize setup and takedown time.

STRATEGIC SUN PLANNING

Anticipate the sun's movement throughout the day when placing shade cloth over containers and raised beds. This strategic placement reduces the need for frequent repositioning, saving valuable time.

 ## How it saves you time down the line

With organization and strategic pre-prep, this streamlined shade cloth pop-up routine is a life saver, protecting your plants from heat stress in minutes. This makes the difference between a struggling garden and one that consistently produces throughout the summer, even in relentless heat, saving the time and hassle of replanting.

Harvest Hustle

05 Minutes

As the mid-season arrives, your garden is starting to burst with fresh, delicious produce. To maximize your harvest in 5 minutes, follow these time-saving tips that will have you hustling through the garden and enjoying the fruits of your labor without breaking a sweat.

ORGANIZE HARVEST TOOLS

Once again, organization is key to the time-saving game, preventing wasted time looking for tools and supplies. Keep baskets, shears, and containers of various sizes within arm's reach to avoid unnecessary backtracking.

TIME-SAVING TIP: Use a harvest apron with pockets for easy access to small tools and no-basket harvesting.

HARVEST DURING IDEAL TIMES

Choose the coolest parts of the day for your harvest hustle, whether it's a morning coffee walk or a family evening garden stroll. This not only makes your harvest more enjoyable, but also enhances the flavor of some veggies when harvested in the cool of the day.

BATCH HARVESTING

Optimize your time with a batch approach—harvesting similar crops at peak ripeness, focusing on one crop each harvest hustle. For instance, dedicate one day to tomatoes, the next day to peppers, the following day to cucumbers, minimizing back and forth trips. Gather multiple similar fruits or veggies in one go, saving time, and allowing for efficient storage.

IN-THE-GARDEN VEGGIE WASH

For crops that need cleaning before storage, rinse off dirt and debris with the hose before bringing them inside, saving you time during meal preparation. Keep a sponge or veggie brush near your outside wash station.

HARVEST BUDDY

Double your time on your harvest hustle by enlisting the help of a friend or family member. This not only increases efficiency, but turns a garden task into a fun experience.

TIME-SAVING TIP: Host a harvest party that includes a 5-minute harvest hustle and then prepare a meal with the bounty each person harvested.

How it saves you time down the line

A 5-minute harvest hustle routine is a game changer, helping you make the most of your hard-earned homegrown veggies and make harvesting more efficient and enjoyable. Customize the frequency of your harvest hustle to align with your garden's readiness to enjoy your veggies at the peak of ripeness and minimize waste. Harvest smarter not harder for more veggies on your plate in less time.

Time-Smart Mid-Season Boost

Busy days shouldn't keep you from nurturing your garden. Here's a 5-minute veggie garden pick-me-up for the middle of the season to keep your garden thriving when time is tight. Focus on one garden area every few days, and you'll have your garden boosted and whipped into shape after just a few sessions. The key here is consistency; a little a few times a week goes a long way to supporting your veggie garden.

SPEED WEED (1 MINUTE)

Grab your garden gloves and do a quick zip through the designated garden area and remove those pesky weeds around your plants. Have a 5-gallon (19-liter) bucket nearby to dispose of the weeds as you go to save time going back later to clean up. This quick patrol not only beautifies your garden, but reduces competition for nutrients, space, and water, giving your veggies the upper hand.

SWIFT SOIL CHECK (1 MINUTE)

Get down and dirty. Stick your finger into the soil up to the second knuckle. Is it too dry or wet? Adjust your watering schedule accordingly. A well-hydrated garden is a happy garden!

RAPID TRIM AND TIDY (1 MINUTE)

Armed with your pruners, spend a minute snipping away at any yellowing or dry leaves. This gives plants better air circulation, keeps disease at bay, and redirects your plants' energy toward healthy growth. Plus, its super-satisfying to see the immediate impact on how fresh and polished your garden looks.

EXPRESS PEST PATROL (1 MINUTE)

A minute of prevention is worth a pound of cure. Early identification and intervention save your veggies from potential damage down the line. Inspect both sides of leaves and stems and look for any unwanted visitors. Spray with a quick hit from the hose or use organic pest control.

FAST-TRACK FEED (1 MINUTE)

In your final minute, grab your favorite liquid fertilizer and give your veggies a quick hit—a nutrient-packed snack to give them a pick-me-up. This is especially beneficial after pruning or after harvesting to boost them into action again.

 ## How it saves you time down the line

In just 5 minutes, you've given your garden a mid-season boost, setting the stage for a thriving garden in the second half of the season. Taking care of small challenges now prevents bigger issues later, saving time and energy. Not to mention more veggies on your plate is always the biggest payoff!

Super-Fast Garden Surplus Savers

Got more veggies than you can handle? In just 5 minutes, quickly preserve, share, or save those extra garden goodies. Here's some quick and simple solutions without breaking a sweat.

LIGHTNING-FAST VEGGIE PRESERVE

Slice cucumbers, bell peppers, carrots, or squash into bite-sized pieces. Toss them into jars, add a pinch of salt and a generous number of fresh herbs, then cover this veggie medley with white vinegar. Put a lid on it, shake, and pop in the fridge. Voilà! In just moments, you've created a quick veggie pickle preserve for salads, sandwiches, or quick tangy healthy snacks that will stay crisp and flavorful for weeks.

QUICK VEGGIE GIFT BAGS

It only takes a few minutes of extra effort to spread the garden love—and your friends will *love* it! Make quick veggie gift bags in under a minute. Personalize a reusable bag with the recipient's name using a permanent marker, then fill it with your garden bounty. Drop it off at a neighbor's house, bring to a friend, or give to a coworker. It's a thoughtful way to share the joy of homegrown produce and brighten someone's day with a burst of garden-fresh goodness.

TURBOCHARGE FREEZER STASH

Turbocharge your freezer stash in minutes! Wash and chop your veggies into bite-sized pieces, lay them on a baking sheet lined with parchment paper, and freeze for a few hours. Once frozen, divide into convenient small portions, transfer to a freezer bag or freezer containers, label, toss back into the freezer, and boom! You've created a stash of frozen vegetables in minutes ready to pull out later for soups, sauces, stir fry, or casseroles.

ON-THE-GO VEGGIE SNACKS

Wash and chop your veggies into snack-sized portions—carrots, zucchini, cucumbers, or cherry tomatoes work well. Toss them in recloseable bags to keep them fresh. Whether you're rushing to a meeting, heading out for a workout, or simply need a quick pick-me-up, these pre-packed snacks mean you can make the most of your fresh produce. Grab one for a quick burst of healthy on-the-go garden-fresh energy, and have an extra on hand to share with a friend.

EFFICIENT MEAL PLANNING/PRESERVING

Plan quick, simple meals around your harvest hustle. Make extra for freezing or giving away. For example, large quantities of grilled or roasted veggies can be flash-frozen on cookie sheets, then transferred into freezer bags for later use. Many veggies, such as tomatoes and peppers, can be frozen whole for later use in sauces and recipes.

 ## How it saves you time down the line

Investing 5 minutes now in preserving, sharing, or freezing surplus veggies, you avoid waste and have a ready supply of fresh veggies for meals and snacks. This quick solution minimizes waste and maximizes enjoyment of your garden's bounty.

Hornworm Hunt

Discovering the dreaded tomato hornworm on your plants can be disheartening. Often appearing later in the season, they have voracious appetites and can quickly wreak havoc on your tomato plants, consuming entire branches overnight. A quick and easy tomato hornworm hunt can save your garden from destruction by these dreaded creatures and ensure your tomatoes thrive.

STEP 1 Gear Up

Grab a pair of garden gloves and a bucket. These are your trusty weapons for the hunt. The gloves make it easier and more pleasant to pluck the hornworms off the plants, and the bucket is for disposing of them.

STEP 2 Scout 'Em Out

Start your hunt by inspecting the upper and lower surfaces of tomato leaves, stems, and branches. Hornworms are masters of disguise, so check each plant carefully for their dark green bodies with white markings and the classic horn.

STEP 3 Follow the Droppings

Look for their telltale signs—they leave behind dark, pellet-like droppings. Follow the droppings to quickly locate their hiding places. If you spot the droppings, the hornworm is likely nearby.

STEP 4 Pluck and Drop

Once you spot a hornworm, quickly pluck it from the plant. They can be quite clingy, grabbing on for dear life, so a firm tug may be needed. If they are stubbornly hanging on and you get squeamish, use a pair of tongs or opt to cut the tomato leaf instead of plucking. Drop the hornworm into the bucket.

STEP 5 Dispose of the Intruders

Once you've collected the intruders, it's time to decide their fate. Despite the destruction they wreak in the garden, they do turn into the beneficial sphinx month. If you're reluctant to harm them, relocate them to a spot away from your garden. Another option is to drop them in a bucket of soapy water or place them in the green waste bin with a few tomato leaves to keep them company until trash day. For severe infestations, spraying with an organic pesticide may be needed.

STEP 6 Rinse and Repeat

Wrap up your hornworm hunt by doing a quick scan of the rest of your garden, as they can also hide on eggplant, peppers, and potatoes. Rinse and repeat steps 1–5. The key is to stay vigilant in the hunt, making this a regular part of your weekly garden routine.

 ## How it saves you time down the line

A 5-minute hornworm hunt now saves extensive damage later down the line. Early detection and removal of these pests not only protects your plants but saves time and effort that would otherwise be spent on extensive pest control measures and promotes a thriving and fruitful tomato crop.

Snappy Spent Plant Cleanup

05 Minutes

In the final stretch of the gardening season, dedicating just 5 minutes to removing old plants does more than just beautify, it significantly impacts the garden's overall health.

DISEASE PREVENTION

Leaving spent plants in the garden creates a haven for pests and diseases, setting the stage for potential problems the next growing season. Clearing them reduces the risk of infections spreading and contributes to pest and disease prevention.

TIME-SAVING TIPS

- **Gather your tools:** Before you start, gather your tools: pruners, gloves, and disposal bins. Having everything at your fingertips saves time and makes the process efficient.

- **Zone-by-Zone Approach:** Instead of attempting to tackle the entire garden at once, break it down into manageable zones. Spend just 5 minutes in one section, cutting spent plants at the base and disposing of them as you go. Systematically move through the garden in several sessions. This approach results in thorough cleanup over time, without feeling overwhelmed.

- **Assess:** Deciding whether to remove plants involves two key factors: appearance and production. Remove those at the end of their life cycle that are brown, yellowing, diseased, or pest infested. Weigh the benefits of potential harvests against the plant's declining condition. Evaluate if replacing it with a new, healthy plant is a better trade off, considering the challenges of nursing an old plant back to health, available space, and the remaining time in the growing season.

- **Set a Timer:** Challenge yourself to complete a zone in 5 minutes. Setting a timer adds as sense of urgency and gives you a sense of accomplishment when you complete a task in the set time.

- **Multi-Tasking for Efficiency:** Efficient gardening always involves multi-tasking! Combine cleanup with other tasks, such as checking for pests, hydration, or harvesting. Make notes for future maintenance or planting sessions.

- **Involve the Family:** Turn plant removal into a family activity. Designate specific tasks or areas to each family member to make the process faster and more enjoyable.

- **Plant Disposal:** Add healthy spent plants to the compost pile. Dispose of diseased or pest-infested plants in the trash or green waste bin to prevent contamination. Keep two buckets at your side and put each plant in the appropriate bucket.

How it saves you time down the line

Late-season spent-plant removal is a time-saving investment. You're preventing disease and pest-infested plants from overwintering, reducing the risk of infestations next season. You'll have a cleaner, more beautiful garden that you can enjoy to the fullest.

5-Minute Late-Summer Garden: Maximizing Harvests in Minimal Time

For the time-strapped gardener, spending just 5 minutes a few times a week in planting a late-summer garden can make sure you have continual harvests before cooler fall weather sets in.

As your current crops wind down, knowing what to plant and when is crucial. Start seeds for your late-summer garden 6–8 weeks before cooler temperatures arrive, and focus on veggies with a short harvest time of 6–8 weeks. Beans, squash, cucumbers, determinate or dwarf tomatoes, and hardy greens such as kale, chard, and collards are ideal for a fresh wave of produce in the late summer.

TIME-EFFICIENT OPTIONS FOR LATE SUMMER GARDEN

STEP 1 **Start from seed outside in six-packs or small containers.**

▶ It's ideal if the garden beds are full while waiting for crops to finish.

▶ Take advantage of warm, late-summer weather for rapid growth, skipping the need to start seeds indoors, without the hassle of setting up grow lights and avoiding the time it takes to harden off seedlings.

▶ Place seedling containers strategically in a location where they get morning sun and afternoon shade to avoid overheating.

▶ Transplant these seedings into vacant spots in garden beds and containers as existing plants reach the ends of their lives.

STEP 2 **Plant seeds directly in garden beds or containers.**

▶ Use your 5-minute planting sessions to directly sow seeds into empty spots.

▶ By the time existing crops are removed at the ends of their lives, seeds will have developed into seedlings and will thrive in the freed-up space and sunlight.

As you grow a late-summer garden year after year, the proactive time-saving benefits become apparent. Starting before your current crops finish prevents supply gaps and avoids the time-consuming process of reviving struggling plants. It's often easier to start anew with fresh plants, especially if spring-planted veggies are reaching the ends of their lives late in summer.

 ## How it saves you time down the line

By investing a few minutes regularly, you'll enjoy a consistent harvest of fresh veggies with minimal maintenance. Planting quick-growing varieties strategically and utilizing all available spaces will reward you with a continuous harvest of homegrown goodness. As the fall chill sets in, the 5-minute sessions pay off as you revel in a thriving garden, and you'll be so glad you took the time to put this garden strategy on autopilot.

05
Minutes

Quick and Simple Cold-Season Cover

As the late season winds down and transitions into colder weather, creating simple cold-season covers is essential to safeguard your warm-weather crops. This extends the growing season and helps you get a few additional harvests from your veggies.

MATERIALS YOU'LL NEED

▶ Bins: Clear plastic bins or storage containers are ideal for covering small containers or small individual plants like beans or dwarf varieties.

▶ Heavy plastic, stakes and clips, and old quilts are suitable for larger plants or garden beds.

STEP 1 **Measure and Cut**
Measure the size of the garden bed or container you want to cover and cut plastic to the appropriate size.

STEP 2 **Create a Frame**
Insert stakes around the garden bed or place evenly around containers to create a stable frame.

STEP 3 **Cover Placement**
Carefully drape the plastic or quilt over the frame, with a few inches (5–7 cm) of overhang to trap warmth effectively.

STEP 4 **Secure Edges**
Use clips, clothespins, bricks, or weights to secure the plastic to the stakes or edges of the garden bed. This prevents air from escaping and keeps the cover in place on windy nights.

STEP 5 **Cover Removal**
Remove covers in the morning if the temperature is above freezing. This helps prevent plants from overheating under the plastic and allows them access to sunlight.

TIME-SAVING TIPS

▶ **Pre-Cut Covers:** Measure and cut covers in advance to avoid last-minute scrambling on cold nights. This pre-prep results in quick and efficient covering when needed.

▶ **Label Containers:** Clearly label which plastic bin corresponds to which plant. This simple step saves time when temperatures drop quickly.

▶ **Store Covers Nearby:** Keep covers stored nearby in a garden shed for quick access. Dual-purpose storage bin covers keep pre-cut plastic, quilts, and clips organized.

How it saves you time down the line

Creating easy cool-weather covers in advance streamlines the process of protecting your veggies. When cool weather or light frost is anticipated, you can quickly cover your veggies, reducing the risk of damage. This not only saves valuable time but also extends your harvests, allowing you to enjoy your garden bounty longer.

05
Minutes

Hassle-Free Herb Propagating

As the gardening season winds down, a hassle-free technique for extending your fresh herb supply into the winter is by propagating, or creating new plants from herb cuttings harvested from your garden. Being proactive now not only saves time and effort starting herbs from seed indoors, but also eliminates the expense of buying short-lived, flavorless herbs at the grocery store.

HASSLE-FREE HERB PROPAGATING

STEP 1 Take Cuttings

▶ Opt for easy-to-root soft-stemmed herbs such as basil and mint. Woodier stem varieties like rosemary, sage, lavender, and thyme take a bit longer to root, but are certainly worth the effort.

▶ Cut a 6–10-inch (15–25.5-cm) stem with new, green growth that has at least four to six leaves.

▶ Strip the bottom few inches (5–7 cm) of leaves from the stem.

STEP 2 Water

▶ Place the stripped end of the stem in a jar of water on a windowsill, away from direct sunlight.

▶ Add a few drops of liquid fertilizer or rooting hormone, such as worm tea, to speed up the rooting process.

▶ Keep a close eye on the water level to prevent drying out, and change the water every 3–5 days to avoid stem rot.

STEP 3 Look for Roots

Keep an eye out for root development. Roots typically emerge on herbs that are easy to propagate, like basil and mint, within 7–10 days, while woodier herbs, like sage, rosemary, and thyme, may take 3–4 weeks. This important step indicates successful propagation and healthy growth of your cuttings.

STEP 4 Plant

▶ Once the cutting has a well-developed root system 1–2 inches (2.5–5 cm) long, transfer it to a small 4-inch (10-cm) container filled with soil, keeping it out of direct sunlight.

▶ When roots are established in the small pot (indicated by a slight resistance when cutting is gently tugged), usually about 2 weeks, transplant into a larger container and place under grow lights.

STEP 5 Harvest

Once your new herb plant is 6–8 inches (15–20 cm) long with new growth, the leaves are ready to harvest and enjoy in garden-fresh recipes.

 ## How it saves you time down the line

Hassle-free herb propagating has many time-saving benefits. You skip the lengthy process of starting herbs from seed indoors and have a fresh supply to harvest sooner. You also save time and money by growing your own fresh herbs indoors without having to purchase herbs at the grocery store. The small time investment pays off, providing you with a convenient, aromatic, flavorful herb supply for the winter months.

Time-Savvy Leaf-to-Mulch

In the off-season, capitalize on the golden opportunity presented by fallen leaves, one of the most valuable free garden resources. In just 5 minutes, collect and shred these leaves to create nutrient-rich mulch for your garden. Here's how to make this a quick and efficient process.

QUICK LEAF COLLECTION

Focus on areas heavy with leaf cover, like corners or under trees, to collect the most leaves in the least time. If your outdoor space lacks leaves, visit a local park. Lay out a tarp and rake the leaves into the tarp for easy transporting and dump into a bin. This not only tidies up the garden but also provides free material for your mulch or compost. Involve the kids and have a leaf collection party, after a round of jumping in the leaf pile.

TIME-SAVING SHREDDING

The goal is to break the leaves down into smaller bits for faster decomposition and so they can be used effectively as mulch. Lay out leaves on the grass and, using a lawn mower with a bagging attachment, run over to shred them. Once the grass catcher bag is full, dump into a collection bin. Alternately, place them in a trash bin and shred with a weed whacker (always wear safety glassed for eye protection).

TIME-SAVING TIP: Pre-bag shredded leaves in manageable sized bags as you collect them for easy transport to where the mulch is needed.

5-MINUTE WINTER MULCH TASKS

Mulching Garden Beds

Spread a 2–3-inch (5–7.5-cm) layer of leaves over empty garden beds to suppress weeds to cover bare soil, prevent erosion, and insulate winter vegetables. Over the winter, leaves break down naturally with the rain and snow, adding organic matter to your garden beds, bringing in the worms and enhancing soil fertility for spring crops.

Compost

Layer shredded leaves into the compost bin along with other compost materials (page 174) for quicker nutrient-rich decomposition. Top your compost with a layer of shredded leaves each time you add new kitchen scraps to fend off fruit flies, gnats, and critters.

Protect Perennials

Shredded leaf mulch acts as a protective blanket for perennial flowers, bulbs, herbs, shrubs, and berry bushes. It insulates roots from cold temperatures and provides slow-release nutrients, boosting spring growth.

Container Insulator

Spread 2–3 inches (5–7.5 cm) of shredded leaves as a top dressing and insulator for overwintering container plants. This will insulate cold-hardy greens to survive freezing temperatures and thrive for winter harvests.

Quick Spread

Keep your bagged leaves handy, so when frost is forecasted, a quick 5-minute dash does the trick to add an extra layer of mulch to the base of your plants for frost protection.

 ## How it saves you time down the line

Leaf collection, shredding, and mulching are a small time investment that reaps big rewards. Utilizing this free natural resource sets the stage for an efficient nutrient boost in the early season. This prepares the spring garden for a productive season with minimal effort on your part.

Microgreens in Minutes

Growing microgreens indoors is a quick and satisfying indoor garden project when the weather is cold and the outdoor garden is dormant. In just 5 minutes, you can set up a flourishing microgreens garden that not only boosts your spirits but also delivers a nutritional punch. Microgreens are harvested at just 2–3 inches (5–7.5 cm) tall, taking up minimal space and making them ideal for an indoor garden.

HOW TO PLANT MICROGREENS

STEP 1 Containers and Soil

▸ Opt for containers that are 4–6 inches (10–15 cm) tall. Microgreens have shallow roots, so there's no need to grow them in a container larger than 1 gallon (4 l).

▸ Use aerated containers such as biodegradable Cow Pots, or in Smart Pots. All containers should have holes for drainage.

▸ Fill containers with pre-moistened potting or seed starting mix, lightly tamping down to eliminate air pockets.

▸ Place containers in shallow drip trays for a neat indoor garden and efficient watering.

STEP 2 Quick Seed Planting

▸ Densely sprinkle quick-growing varieties like mustard greens, arugula, radishes, and kale in different containers. No need to space seeds evenly.

▸ Accelerate germination by lightly covering seeds with soil (too much soil will slow germination) and insert plant labels.

▸ Optimize growth by placing under grow lights with an 18 hours on/6 hours off schedule or place them in a sunny windowsill.

EFFICIENT WATERING TECHNIQUES

▸ Before germination, keep soil moist by misting with a spray bottle.

▸ After germination, bottom-watering streamlines the watering process.

▸ Pour several inches (5–7 cm) of water in the bottom of the drip trays, allowing water to absorb from the bottom.

▸ Pour off water once the top of the soil is a dark brown color, indicating adequate moisture.

▸ Bottom-watering allows for more complete water absorption, reducing the frequency of watering and saving time.

HARVEST AND ENJOY

▸ Harvest microgreens in 7–10 days when they are 2–4 inches (5–10 cm) tall.

▸ Snip a small handful at the base as needed.

▸ Enjoy microgreens in salads, sandwiches, wraps, smoothies, or as delicious toppings for soup, chicken, or fish.

 ## How it saves you time down the line

The rapid growth of microgreens allows you to enjoy nutrient-dense greens indoors within a week even when the outdoor garden is dormant for the winter. Microgreens are a hassle-free way to keep your green thumb happy year-round, making it easy to maintain a healthy lifestyle with minimal time investment and space commitments.

Lightning-Fast Grow Light TLC

In the off season, efficient grow light maintenance is essential for optimal performance and a smooth transition into the early growing season, or for successful indoor gardening during winter. Spending just 5 minutes on each step over a few weeks makes it lightning-quick and makes sure your garden is ready for successful seed starting.

STEP 1 Set Up and Check Bulbs

▶ If grow lights aren't set up for indoor gardening, set up one small grow light area in each 5-minute session. Turn on each bulb and look for dimming or discoloration, signs of wear and tear, and loss of intensity.

▶ Replace bulbs if needed to ensure the optimal light spectrum for healthy plant growth.

STEP 2 Clean

▶ Turn off grow lights and let them cool. Conduct a quick visual check for accumulated dust or debris.

▶ Gently clean bulbs and top of the lights with a soft, dry cloth.

STEP 3 Time-Saving Timer Installation

▶ Purchase a timer for a streamlined grow light schedule if you haven't already. Consistency in light exposure is crucial for seedling growth.

▶ A timer eliminates the need to manually turn on/off lights and prevents forgetting.

▶ Test your timer with your grow lights to make sure it works. My preferred schedule for seedings is 18 hours on/6 hours off.

STEP 4 Grow Light Station Check

▶ Check the shelves your grow lights are attached to as part of your lightning-fast routine.

▶ Make sure they are in good working order, fix any damaged shelves, or adjust shelves for different plant heights.

▶ Have a shallower shelf for seeds that are just starting, and a taller shelf with more room in between for growing plants.

▶ Collect wood blocks or shoeboxes to adjust the height of growing plants under the grow lights.

 ## How it saves you time down the line

Completing a step of this lightning-fast routine each week during the off season pays off in the long run. Consistent light exposure and hassle-free timers result in healthy plant development and reduces the likelihood of leggy or stunted seedlings. You'll be set up for successful seed starting, leading to strong, healthy seedlings and a more productive garden when the growing season starts.

Speedy Supply Check

Spending 5 minutes a few times a week during the dormant season on a speedy garden supply check is a game-changer for economical and efficient gardening. Here's a quick guide to make this routine streamlined to enjoy long-term time and money-saving benefits.

STEP 1 Inventory Assessment

▶ Take inventory of your garden supplies, including soil, fertilizer, seeds, seed-starting supplies, containers, and tools.

▶ Note any broken items that need replacement.

STEP 2 Economic Planning

▶ Use your inventory assessment to identify what's lacking and make a wish list, prioritizing items in order of need.

▶ Being prepared with a list means you can take advantage of off-season sales and only buy what you need.

▶ This strategy eliminates impulsive or "have to" purchases and last-minute trips to busy garden centers during the peak season.

STEP 3 Online Browsing

▶ Dedicate a few 5-minute sessions (set a timer to stay focused) to peruse garden websites and online seed shops.

▶ Set up an app with sale alerts or coupon notifications for your favorite suppliers.

▶ Stick to your garden budget by only purchasing items on sale and those on your list.

STEP 4 Off-Season Clearance

▶ Take advantage of off-season sales offered by many garden centers to clear out extra inventory.

▶ Stock up on essentials at the lowest prices during this period.

▶ Opt for economical slow shipping options, since your garden needs are not urgent during the off season.

ALL STAR GOURMET
LETTUCE MIX
SEEDS

SUPPLY INVENTORY:
Potting mix – 3 bags
Seed-starting containers – 10
Plant tags – 25
Coco coir – 5 small bricks
Worm castings – 2 bags
Granular fertilizer – 1 small bag
NEED:
Bin to mix seed-starting mix
Clamp light
Grow light bulbs
Vermiculite
Trowel – broken
SEEDS:
Need to buy:
– tomatoes
– beans
– sunflowers
– zinnias

 ## How it saves you time down the line

This 5-minute supply check during the off-season lays the groundwork for a smoothly run, efficient gardening season. Through strategic planning, capitalizing on sales, and avoiding last-minute purchases, you not only save valuable time and money, but also make sure you have the necessary supplies to plant and nurture a productive garden. Your garden will thrive, and you can enjoy the stress-free satisfaction of knowing your garden journey is budget friendly.

Streamlined Seed Organization

Organizing seeds can be a breeze with just 5 minutes a day. Dedicate a week during the off season to spend 5 minutes a day to transform the chaos of seed packets into a well-organized stash, saving time and stress in the planting season. Here's how to set up your streamlined seed storage system.

STEP 1 **Gather Your Seeds**
Collect seed packets or saved seeds in various locations. Having everything in one place sets the stage for quick and efficient organization.

STEP 2 **Sort by Planting Season and Alphabetically**
Categorize seeds by planting season: spring, summer, fall, cool season, and warm season. Alphabetize seed packets within each season to speed seed retrieval according to your planting schedule.

STEP 3 **Store in Clear Containers**
Opt for clear containers like plastic shoe bins, binders with photo sleeves, or colorful photo boxes of uniform sizes. For saved garden seeds, use mason or spice jars with lids or resealable bags. Clear containers facilitate visibility and quick access, maximizing your gardening minutes. Store containers in a cool, dry, dark place to keep seeds viable for as long as possible

STEP 4 **Label Clearly**
Label each seed envelope or container with the vegetable type and planting season. This is your shortcut to the right seeds, significantly reducing search time. Use tabs or labeled dividers, or make dividers from card stock or sticky notes for your storage containers to further subdivide for quick locating.

STEP 5 **Create Quick Reference List**
Keep a notepad or digital document handy to create a quick reference list of your seed inventory, noting the quantity of each seed variety. This list becomes a time-saving tool for garden layout and seed ordering, preventing duplicate purchases.

Warm Weather A-L

Warm Weather R-Z

Cool Weather L-Z

Watermelon

Zinnias

Tomatoes

Flowers and Herbs

CaliKim's Pepper Seed Collection

TOM THUMB PEA
SEEDS

⏱ How it saves you time down the line

Efficient seed organization provides quick access to seeds based on planting season and types. Clear storage and uniform-sized storage containers and labeling not only minimizes search time, but also allows you to work smarter, not harder, and leads to a streamlined planting process. You'll be able to plant more seeds in less time, for a less-stress gardening experience that is a pleasure rather than a chore.

08
Minutes

CHAPTER THREE

Powerful 8-Minute Accomplishments

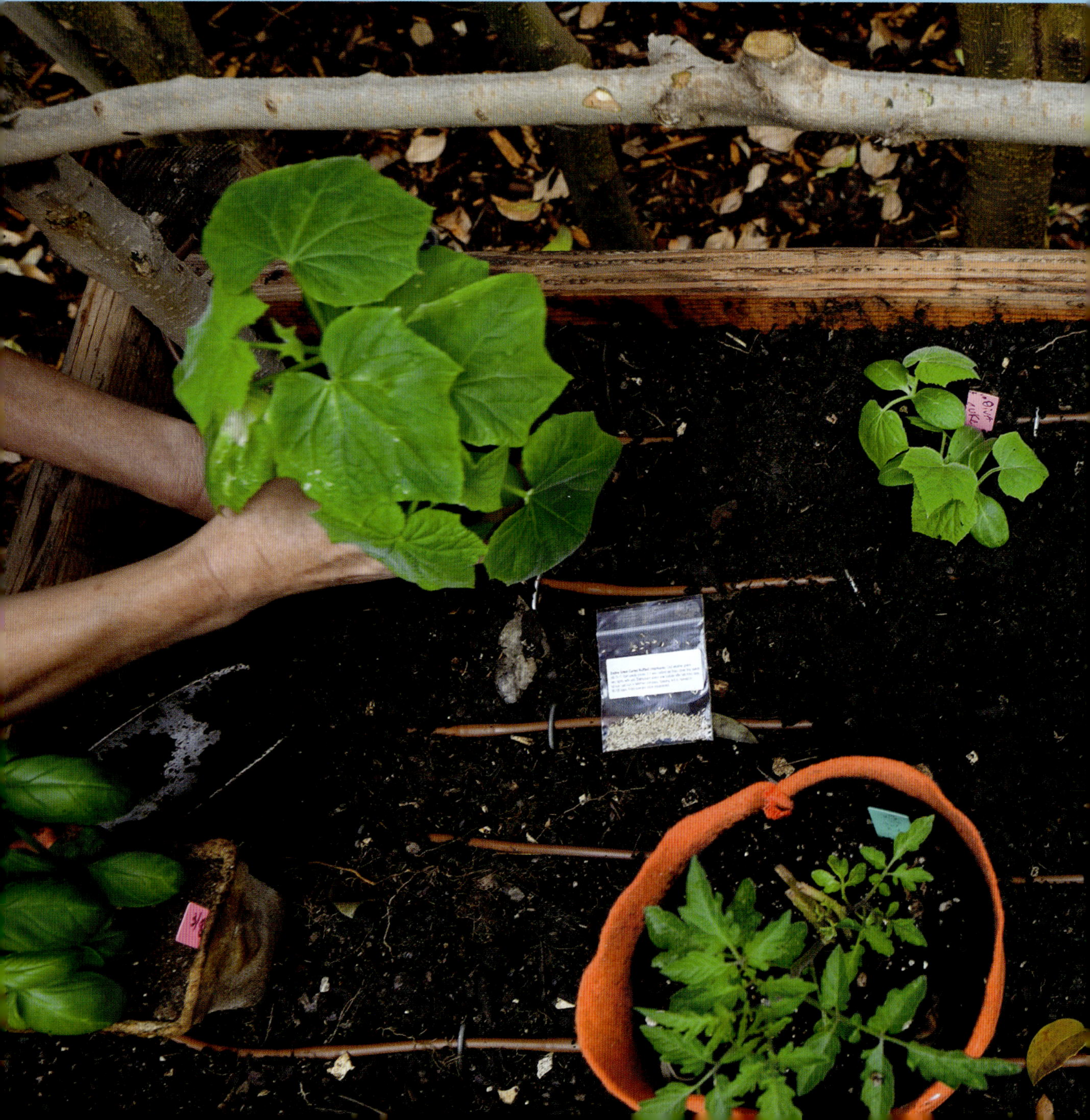

What to Do When 8 Minutes Is All You Have

POWER TASKS

In our fast-paced lives, finding time for gardening can be tough, but with Powerful 8-Minute Accomplishments, you can make significant progress in short, manageable bursts. Whether you're a parent, a busy college student, or a working mom, you can find 8 minutes to fit into your day. You'll learn how and when to seamlessly integrate these quick and powerful 8-minute tasks into your day, as opposed to when to choose a 3- or 5-minute task.

GREATER DEPTH

An 8-minute task allows more time for tasks that require time and attention, such as planting a small flower bed, pruning multiple plants, or building a simple trellis. In contrast, a 3- or 5-minute task is ideal for quick maintenance tasks like deadheading a few flowers, watering container plants, or conducting quick pest inspections.

PROJECT COMPLETION

An 8-minute time frame allows for substantial progress on ongoing projects that have multiple steps. For example, setting up a small raised bed may require soil preparation, arranging the plants, watering and mulching—taking multiple 8-minute sessions to complete.

ATTENTION TO DETAIL

With 8 minutes, you can pay more attention to detail than is needed to complete the 3-minute tasks, which are geared toward quick inspections, light maintenance, or addressing minor issues.

SEASONAL TASKS

Eight-minute tasks are ideal for seasonal tasks that might require a bit more time, such as preparing garden beds for the winter or setting up winter or shade cloth covers. In contrast, 3-minute tasks are better suited for quick, routine, year-round maintenance that needs to be done frequently.

How to Fit 8-Minute Tasks into Your Day

MORNING BOOST

Start your day with a burst of garden energy and get a light workout. Carry buckets of soil to prep for projects, or complete a pruning session to kick-start the day.

SCHEDULE A GARDEN BREAK

Take a break from indoor tasks to get outdoors to enjoy making progress on a garden project. It sets a positive tone for the day and helps clear your mind for greater focus. Tending to plants, sowing seeds, or tidying tools provides a refreshing mental break.

POST-WORK GARDEN THERAPY

Decompress after work by heading to the garden to fill a container, hand watering, or appreciating the beauty of the garden before preparing for dinner.

MULTI-TASK MAGIC

Combining a task with another activity, such as listening to a podcast while watering or thinning seedlings during a phone call, is a smart way to accomplish more in less time.

WEEKEND GARDEN SPRINT

Invest 8 minutes on a Saturday or Sunday to tackle a specific garden task such as seed starting, pest control, or collecting materials for a raised bed.

Dedicating 8 minutes here and there allows you to break down larger tasks into manageable segments, providing a sense of accomplishment even with completion of small tasks.

Seed-Starting Saturdays and Sundays

Seed-Starting Saturdays and Sundays are a powerful, simple routine that make this task manageable. Investing just 8 minutes each weekend day during the early season makes caring for your seedlings over time manageable, avoiding the overwhelm of starting too many at once. Utilizing your designated seed-starting station (see page 136) streamlines the entire process, helping you kick-start your spring garden effortlessly in just 16 minutes a week.

Here's a detailed breakdown of the Seed-Starting Saturdays and Sundays routine:

▶ **Set a timer for 8 minutes.** This limit keeps you focused. Seed starting will feel more like an enjoyable ritual than a chore.

▶ **Designate crop types.** Assigning one type of crop to each seed starting session keeps you organized and gives you a wide range of crops in your garden. For example: Dedicate Saturday to starting tomatoes, and Sunday to starting cucumbers. Introduce two new crop types each week to keep things interesting and your garden diverse.

▶ **Seed starting steps.** Follow a routine to maximize efficiency. First, moisten the seed starting mix, then fill the six-packs or seed starting containers, place them in drip trays, plant the seeds, insert the plant tags, spray containers with water, and finally, place the trays under grow lights.

▶ **Rinse and repeat.** Consistency is key to this routine. Repeating these short sessions not only builds muscle memory, but also makes you faster and more efficient with each repetition.

How it saves you time down the line

Consistent 8-minute Seed-Starting Saturday and Sunday sessions are powerful and will bring you closer to the garden of your dreams, in less time than you imagined. A diverse variety of seedlings will be growing and flourishing indoors, giving you an early advantage to your spring garden. Your homegrown transplants will be ready for planting outside when weather permits, which means earlier harvesting of fresh, flavorful veggies from your own garden grocery store.

08
Minutes

Time-Efficient Seedling Thinning

When it comes to growing healthy indoor seedlings, thinning is a crucial step that is often overlooked. When seedlings are sown too closely together, they become overcrowded and compete for resources like air, nutrients, and light, hindering their growth. However, thinning them out can make a significant difference, giving the strongest seedlings more space for growth. Here are step-by-step instructions to thin seedlings effectively.

STEP 1 Early Identification
Identify which seedlings need thinning by looking for signs such as stunted growth, yellowing leaves, or a lack of growth. This early identification helps you streamline the process from the get-go.

STEP 2 Choose the Strongest Seedlings
Chose the strongest seedlings with vibrant color, sturdy stems, and well-developed leaves. These are the ones that have the best chance of thriving when given more space.

STEP 3 Snip—Don't Pull
Use sharp precision pruners or small scissors to snip the weaker seedlings at soil level. Avoid pulling them out to minimize root disturbance. Clean up any dead or yellowing leaves to improve the health of the remaining seedlings.

STEP 4 Spacing Is Key
Leave enough space between remaining seedlings to allow for air circulation, light, and additional growth. Refer to specific spacing requirements for each plant variety.

STEP 5 Nutrient Boost
After thinning, give remaining seedlings a nutrient boost with an organic liquid fertilizer or a gentle growth hormone like worm tea. This ensures the plants receive essential nutrients for additional growth.

STEP 6 Enjoy the Thinnings
Use thinned greens, carrots, and herbs in a salad or as microgreens to top soups or in sandwiches or wraps.

STEP 7 Regular Monitoring
Incorporate thinning into your routine as the seedlings continue to grow. This allows you to address overcrowding right away, and helps your seedlings continue to grow into thriving and healthy plants.

 ## How it saves you time down the line

Thinning indoor seedlings may seem like an extra step, but it's a time-saving investment. Providing optimal spacing and removing weaker seedlings early on prevents stunted growth due to overcrowding. This proactive approach results in healthier, more resilient plants, which pays off with thriving plants down the line.

Plant Protection Powerhouse

A thriving garden involves not just growing plants, but also safeguarding them from critters and pests. Combining three simple yet effective techniques creates a plant protection powerhouse for your garden. Investing a few brief sessions in the early season results in healthier, resilient plants and more veggies to harvest.

TECHNIQUE #1: COMPANION PLANTING

Strategically placing pest-repelling or beneficial-insect-attracting plants near vulnerable vegetables is a game-changing technique known as companion planting. My favorites, calendula and nasturtiums, are easy to plant from seed. Their strong scents not only repel pests but also attract beneficial insects, like ladybugs, hoverflies, and lacewings, that are natural pest predators. These flowers not only provide protection, but also add a burst of color to your garden.

TIME-SAVING TIP: Keep calendula and nasturtium seeds handy in a basket by the backdoor for quick and easy planting in early spring and fall.

TECHNIQUE #2: ROW COVERS

Physical barriers in the form of row covers shield your veggies from insects and small critters, like birds, squirrels, and rabbits. Tulle, my preferred material, is an inexpensive netting-type material that comes in many different colors. It allows in sunlight and airflow while keeping out pests and critters. For UV protection, spray tulle with sunscreen before installing. Install row covers at planting time to prevent bugs from laying eggs under the tulle.

TIME-SAVING TIP: Cut the tulle a little larger than the garden bed or container, and secure it around the edges with bricks or clips for easy installation and removal.

TECHNIQUE #3: NATURAL REPELLENTS

Create a protective barrier around your plants by making a homemade spray with ingredients like neem, garlic, or peppermint oil, mixed with water. Apply to plants to deter pests and critters.

TIME-SAVING TIP: For convenience, keep an organic, ready-to-use natural pest repellent product in your garden shed.

 ## How it saves you time down the line

Implementing these techniques from the start not only protects your garden, but also saves time by preventing pest and critter issues. You'll enjoy a thriving garden and have many more harvests free from the constant frustration and struggles of seeing your plants damaged.

Hassle-Free Container/ Raised-Bed Soil Refresh

Growing veggies in raised beds and containers is convenient, but soil loses nutrients over time. Revitalizing the soil in your raised beds and containers, rather than replacing it, saves time and money. Complete this hassle-free process in one to three 8-minute sessions, depending on your garden's size. Take it step-by-step for gradual progress without overwhelming yourself.

No need to dump out soil from your containers for this process. Simply top off or replace one-third of the old soil to make the process time-efficient and cost-effective.

STEP 1 Remove Old Plants

Clear out old plants to make room for fresh, productive ones. Take out roots, if needed, especially in containers where space is limited. Remove about one-third of the existing soil and set it aside for later use as filler for raised beds or containers. If soil has settled, there's no need to remove one-third, just top it off.

TIME-SAVING TIP: Instead of completely removing old plants, cut them at soil level and leave roots in to decompose in place, adding organic matter to the soil over time.

STEP 2 Add Compost or Fresh Potting Mix

Whether you're removing one-third of the soil or topping it off, add 2–3 inches (5–7.5 cm) of compost and work it into the soil for a quick and easy nutrient boost to keep your soil fertile. This can be homemade, or storebought for a simple but effective amendment.

TIME-SAVING TIP: Purchase compost in bulk at local landscape companies and store in trash cans for easy access.

STEP 3 Add Worm Castings—Natural Fertilizer

Boost life in tired soil with worm castings, an excellent natural fertilizer. Add a few handfuls to containers and work into the soil, or apply ¼ inch (0.5 cm) to raised beds. This provides essential nutrients, beneficial bacteria, and microbes, improving soil structure and protecting plants

from disease. Worm eggs in the castings hatch into more worms, leaving behind a steady supply of nutrients.

TIME-SAVING TIP: Purchase bagged worm castings from a local supplier to eliminate the need to maintain your own worm farm.

STEP 4 Add Organic Granular Fertilizer with Balanced NPK

For added nutrients, spread an organic granular fertilizer with a balanced NPK (nitrogen, phosphorous, potassium) throughout the garden bed or directly into a planting hole if on a budget. This ensures a well-rounded supply of essential nutrients for the next round of veggies.

TIME-SAVING TIP: Store fertilizer in repurposed coffee cans in a storage bin near your garden for easy and time-efficient application.

STEP 5 Water

Water until the soil is evenly moist, not waterlogged, to allow the amendments to absorb into the soil. Adequate watering prevents nutrient run-off and supports the overall health of your refreshed and revitalized soil.

TIME-SAVING TIP: Invest in a soaker hose or drip irrigation to reduce the time spent manually watering.

How it saves you time down the line

Regular seasonal soil refreshes your soil's fertility, avoiding expensive and time-consuming soil replacements. This keeps your garden producing veggies and gives you more time to enjoy your garden without the backbreaking labor of dumping out large amounts of soil. This makes gardening hassle-free and enjoyable.

Bamboo Blitz Trellis

The Bamboo Blitz Trellis is an easy-to-assemble, tool-free garden beauty that brings both charm and efficiency to your outdoor space. This simple-yet-stylish trellis requires no tools or wood cutting and can be made in just 5 minutes, transforming your garden and expanding your growing space. It uses eco-friendly materials of bamboo and thick twine to provide a sturdy solution for your climbing plants, while adding a touch of natural beauty.

MATERIALS YOU'LL NEED

▶ Three 6–8-foot (2–2.5-meter) bamboo poles (at least 1 inch [2.5 cm] in diameter, thicker for added sturdiness)

▶ Thick, sturdy twine

STEP 1 Assemble the Bamboo Poles
Begin by tying the three bamboo poles together at the top with twine or wire. This creates a stable structure for your trellis so it will stand tall and strong in your garden bed.

STEP 2 Wrap with Twine
Wrap the twine tightly around the bamboo poles to secure them together. Tie off the end of the twine to keep them in place.

STEP 3 Trellis Installation
Place the trellis in your garden bed or container, and fan out the bottom of the poles to form a teepee shape. Press into the soil for stability, and adjust the width based on the location and the plants you're supporting.

TIME-SAVING TIPS:

▶ Pre-cut several lengths of twine before starting to assemble to streamline the wrapping process without interruptions.

▶ Create multiple trellises at once to save time by batching the process and providing uniformity and visual appeal to your garden. This saves valuable planting space.

▶ Prepare garden bed/container soil in advance of making the trellis. This allows for immediate planting once the trellis is in place.

 ## How it saves you time down the line

The Bamboo Blitz Trellis is a quick 5-minute assembly, perfect for time-challenged gardeners. The vertical setup makes veggies more visible and easier to harvest, maintaining their health by keeping them off the ground. This efficient design allows you to enjoy more veggies in less space, maximizing visual appeal and growing space.

Herb-Drying Hustle

The mid-season abundance of herbs is here. Maximizing your harvests is key to enjoying the flavor of your herbs year-round. Preserving herbs doesn't have to be a time-consuming task. Here are three quick and simple methods to dry your herbs efficiently (no dehydrator needed) that take 8 minutes or less of prep time. This way you have a winter stocked with delightful flavorful herbs to add to your recipes!

Herbs are best for drying when harvested before flowering and have the most potency if harvested in the morning when oils are the most concentrated. Wash and pat dry, or use a salad spinner to remove as much moisture as possible before proceeding with your chosen drying method.

AIR-DRYING—THE CLASSIC

- ▶ Tie small bunches together with twine.
- ▶ Suspend the bundles upside down in a warm, well-ventilated area.
- ▶ In 1–2 weeks, your herbs will dry and be ready for storage.

BENEFITS: Minimal effort, and natural air circulation maintains herb quality.

MICROWAVE MAGIC

- ▶ Lay a single layer of herbs on a paper towel on the microwave turntable.
- ▶ Microwave in 15–30 second increments, flipping the leaves over for even drying.
- ▶ Smaller, tender herbs like thyme and tarragon may dry in as little as 1 minute; larger herbs like basil, parsley, and oregano will take longer.
- ▶ Allow herbs to cool, checking for crispness.

BENEFITS: Ultra-fast drying is convenient and perfect for small batches, though quick drying may slightly lessen flavor.

OVEN DRYING

- ▶ **Preheat:** Set your oven to the lowest temperature (around 180°F [82°C]) and leave the door ajar while drying.
- ▶ **Dry herbs:** Lay herbs on a baking sheet on parchment paper in a single layer.
- ▶ **Bake and monitor:** Keep a close eye on the herbs, as drying times vary—usually approximately 1–3 hours. If leaves crumble easily, they're done.

BENEFITS: Drying with even, low heat is ideal for large batches and preserves the flavor more than microwaving.

STORAGE

Store dried herbs in airtight containers away from light and heat to maintain flavor and potency.

 ## How it saves you time down the line

Investing a few minutes in mid-season for herb drying streamlines meal preparations. Quickly reach for dried herbs instead of chopping fresh herbs each time you cook. Enjoy a steady supply throughout the winter, eliminating the need to purchase expensive herbs at the grocery store, and make the most of your fresh herbs year-round.

Summer Salad Station

As the summer heat soars, say goodbye to our cool-weather lettuces and hello to the heat-loving greens of a summer salad station! This is a quick and dynamic 8-minute project that grows salad in one contained garden space to save harvest time and guarantees crisp, delicious salads all summer.

TIME-SAVING TIP
Grow your Summer Salad Station in a compact space, like a pop-up 3-foot (1-meter) Smart Pot raised bed for growing multiple greens. Unfold, fill, and plant—an instant garden. This not only saves you time, but also is a budget-friendly option for busy gardeners.

STEP 1 Clear the Stage
As the cool-weather greens signal the end by bolting, flowering, and going to seed, remove them to make room for our heat lovers.

TIME-SAVING TIP: Cut the plants at soil level for quick removal, leaving the roots in to decompose in place to enrich the soil. Compost the diseased and pest-free plants.

STEP 2 Prepare the Soil
Whether you're planting in a raised bed or a container, success begins with soil prep. Refresh by adding a few inches (5–7 cm) of compost, worm castings, and organic granular fertilizer.

TIME-SAVING TIP: Moistening the soil as you refresh provides immediate hydration for your greens once they're planted, kickstarting their growth.

STEP 3 Plant the Stars
Plant colorful, heat-loving power greens—kale varieties for a burst of color, rainbow chard for vibrancy, red-veined sorrel for a touch of elegance and a citrusy kick, the striking Ruby Red Orach for a pop of red-purple leaves, and New Zealand spinach for a heat-loving spiller that maximizes space.

TIME-SAVING TIP: Start seeds indoors 6 weeks before planting or purchase transplants for a speedy planting.

STEP 4 Nurture the Salad Station
For a thriving salad station, consistent watering and feeding are vital. Deep soak for adequate hydration and feed twice monthly with a liquid organic fertilizer for optimal leaf growth.

TIME-SAVING TIP: Combine watering and fertilizing to save time by adding liquid fertilizer to your watering can during a regular watering session.

STEP 5 Mulch for Heat Protection
Protect your greens from overheating with a 2–3-inch (5–7.5-cm) layer of shredded leaf mulch. This protective barrier minimizes water evaporation and shields roots from the sun's intensity.

TIME-SAVING TIP: Collect leaves in the fall (page 82) and store near your garden beds for quick mulch touch-ups.

STEP 6 Harvest for Steady Supply
Harvest with the cut-and-come-again method by cutting just the outer leaves at 4–6 inches (10–15 cm). Leave the inner leaves to grow to have a perpetual salad bar producing flavorful greens for countless meals so you can grow more food in less space and time.

TIME-SAVING TIP: Harvesting regularly encourages continual production, making the most of your Summer Salad Station.

 How it saves you time down the line

Maximize your time and harvest an abundance of fresh, tasty salads in one small space with a minimal time investment. Utilize the cut-and-come-again technique, allowing you to make the most of your efforts and grow a continuous supply of homegrown salads in less space and time.

Smart and Savvy Trellis Inspection

In the heart of the summer, when your garden is in full swing, take just 8 minutes for a quick trellis inspection to avert disaster and ensure the stability of your vertical growing plants. This smart and savvy tip is a preventive measure so that heavy or overgrown plants don't topple your trellises. This keeps your garden organized and your precious crops safe. Inspecting only 1–2 trellises per session keeps you time-efficient and prevents the task from feeling overwhelming.

STEP 1 **Set a Timer and Gather Your Tools**
Optimize efficiency by setting an 8-minute timer on your phone. Grab gloves, pruners, and ties or supports. Having everything ready before starting will streamline the process.

STEP 2 **Assess Vertical Plants**
Inspect each plant on the trellis for signs of overgrowth, tangled vines, or heavy fruit that might be weighing down the structure. Pay attention to any bending or leaning of the trellis itself.

STEP 3 **Prune and Tie as Needed**
Trim excess to reduce unnecessary weight to the trellis. Remove dead or yellowing leaves. For sprawling vines, guide them back onto the trellis, tying up as you go to secure plants into an upright growth pattern.

STEP 4 **Reinforce Weak Points**
Reinforce weak points by tightening screws, securing joints with cable ties, or adding additional supports like T-posts or sturdy stakes. Taking time to do this now prevents potential trellis topples later in the season.

STEP 5 **Plan for Continued Growth**
Anticipate the weight of future growth and make adjustments with added supports to accommodate growth and to ensure the trellis remains sturdy throughout the entire season.

 ## How it saves you time down the line

This quick trellis reinforcement not only prevents potential disasters, but also contributes to the overall health and productivity of your garden. It's a smart and savvy investment that pays big dividends in the form of a well-organized and thriving garden.

Effortless Edible Flower Harvest

Growing edible flowers not only adds vibrant colors to your garden, but also brings culinary delight! By dedicating a few minutes each week to flower harvesting during the mid-season bloom, you'll not only have more blooms to enjoy and create recipes with, but willalso create a pollinator's paradise.

TIME-SAVING TIPS

Maximize your harvest time with strategic planting. Growing wide swaths of flowers in a concentrated area not only makes harvesting efficient, but also entices pollinators. Plant easy-to-grow flowers that reseed easily with a long bloom time in their preferred seasons. Options for cool-weather lovers include nasturtiums, calendulas, chamomile, pansies, and violas. Sun worshipers like zinnias, marigolds, cosmos, and borage grow well in warmer summer temperatures.

Keep essential tools, like pruners for precise snipping, conveniently stuck in soil in a nearby container for quick access.

SIMPLE HARVESTING TECHNIQUE

Snip flowers with sturdy stem blooms at the base of the stem above a set of leaves. Cut fully open flowers to optimize flavor and longevity, avoiding those that are starting to dry. Delicate flowers like violas, borage, and chamomile can be picked with your fingers. Cut different varieties—the sky is the limit in the kitchen!

EASY USES FOR FLOWER HARVESTS

▶ The most rewarding use for flower harvests is to put cut flowers in a vase or jar to enjoy a bit of the garden inside, or to share with a friend.

▶ Transform your edible flower harvest into lovely culinary treats by topping salads with nasturtiums, pansies, or calendula petals.

▶ Steep in hot water for a soothing and beautiful cup of tea.

▶ Garnish beverages with zinnias and marigolds. Freeze borage blooms in ice cubes for a visually stunning addition to drinks.

▶ Create delightful spreads and dips by adding to cream cheese or butter.

▶ Crystalize flowers by dipping in egg white and sugar for a show-stopper on top of desserts.

 ## How it saves you time down the line

Consistent flower harvesting not only gives you a colorful array of blooms you can use in the kitchen and as cut flowers, but saves time in the long run. Snipping spent flowers redirects the plant's energy into producing more blooms, creating a lush and vibrant garden and a pollinator-friendly environment, which translates into an abundance of veggies. Harvesting flowers while appreciating the colors and aromas is very therapeutic, reduces stress, and contributes to an enjoyable gardening experience.

Efficient Container Feeding

08 Minutes

In the heart of gardening season, keeping your container plants thriving and producing doesn't have to be a time-consuming task. By implementing a streamlined and consistent fertilizing routine, you can maximize your yields while minimizing effort, so you have lots to harvest throughout the season.

WHY CONTAINERS NEED MORE ATTENTION

Container plants require more frequent fertilizing and watering compared to those in raised beds and in-ground gardens. The faster drainage in containers means that water and nutrients leach out quicker.

HOW TO WATER CONTAINERS

STEP 1 Check Soil Moisture

Before watering, always check the moisture level of the soil. Insert your finger into the soil up to the second knuckle. If it feels dry, it's time to water. If it's still moist, hold off for another day or two.

STEP 2 Water Deeply and Thoroughly

When it's time to water, make sure to water deeply until the water runs out the bottom of the container. Use a watering can with a narrow spout or a drip irrigation system to deliver water to the base of the plants. Water until you see excess water draining from the bottom of the container, indicating the soil is saturated.

STEP 3 Mulch and Watering Schedule

Apply a layer of mulch, such as shredded leaves or straw, on top of the soil to help retain moisture. Establish a watering schedule based on the needs of your plants and the weather conditions. Water early in the morning or late in the afternoon to minimize water loss due to evaporation.

TIME-SAVING TIPS

▶ **Use the Right Tools:** Invest in a hose-end sprayer that automatically mixes water and fertilizer. This eliminates the need for manual mixing in a watering can, saving valuable time.

▶ **Set Reminders:** Program your phone to remind you weekly to fertilize your containers. Consistency is key. Phone reminders make sure you don't forget feedings.

▶ **Keep Supplies Handy:** Place a bag of worm castings or compost conveniently near your containers. A quick handful as you pass by provides a quick nutrient boost without disrupting your daily routine.

▶ **Concentrate on Zones:** Divide your garden into four sections and fertilize one area each week. Aim to complete the task in just 8 minutes, setting your timer for each session. By the end of the month, you've covered the entire container garden, which will be well-fed and happy, and you've created a manageable routine.

RINSE AND REPEAT

Repeat the same routine each month. This consistency ensures your plants receive the nutrients they need on a regular basis.

How it saves you time down the line

Efficient container fertilizing is more than just spending time on a routine, it's an investment in the beauty and productivity of your container garden. With the right tools and a consistent schedule, you'll enjoy abundant harvests throughout the season. This approach lets you work smarter, not harder, eliminating the need to correct nutrient deficiencies, and fill your harvest basket time after time from your container garden.

Hardy Greens Smart Pruning Strategy

In the Summer Salad Station (page 106), hardy greens like kale and chard thrive, even in the summer heat. Late in the season, as these sturdy plants mature, they may show signs of bolting. Instead of pulling them out, implement a smart pruning strategy to extend your harvests and make your gardening more time-efficient.

DON'T PULL IT—PRUNE IT!
BENEFITS OF PRUNING OVER PULLING

EXTENDED HARVESTS: Pruning allows you to enjoy a continuous harvest by promoting fresh, new growth at the base of the plant. Instead of starting over with new seedlings each planting season, you're maximizing the potential of the same plant.

YEAR-ROUND GROWTH IN FROST FREE CLIMATES: In milder climates, strategic pruning allows hardy greens to last for several years. This not only saves time, but also allows you to enjoy harvests throughout the year.

COLD AND HEAT TOLERANCE: Kale, chard, and other hardy greens are known for their extreme heat and cold tolerance with prolific leaf growth. With strategic pruning, they become valuable assets both in the garden and on your dinner plate, providing delicious, nutrient-dense food not only in the summer, but also into the winter months, even in climates with harsh winters.

Time-Saving Pruning Tips

STRATEGIC SNIPPING: When the plant bolts or the stem becomes thick and tough, grab your pruners and trim the plant down to a 2–3-inch (5–7.5-cm) stem. This may seem drastic, but it actually stimulates fresh new growth from the base.

STUBBORN STEMS: For exceptionally thick and stubborn stems, use a heavy-duty tool like a lopper for a clean cut. Remove the stem and compost disease- and pest-free leaves. While some of the leaves may still be edible, note that bolting alters the flavor, and the leaves may be best for composting rather than on your dinner plate.

REGULAR PRUNING: Incorporate pruning into your regular garden maintenance routine. This not only manages the size of the plants, but also prevents overcrowding, for a healthier and more productive garden.

 ## How it saves you time down the line

By avoiding replanting the hardy greens every season, you reclaim valuable time. You can redirect this time toward other garden tasks, or spend quality minutes with family and friends. The result? More to harvest that sustains your family, while having surplus to share the joy with friends.

Pronto Pepper Overwinter

Overwintering your favorite and most productive pepper plants is a simple, rewarding process for a quicker pepper harvest in the spring. Peppers, typically labeled as annuals, have the potential to be grown as tender perennials as long as a few steps are taken to keep them alive over the winter, a technique known as overwintering. Here's your quick, simple four-step guide to overwinter your favorite pepper plant—pronto!

STEP 1 **Harvest Peppers**
Begin the overwintering process when pepper production begins to decline, when nighttime temperatures hit the 40°F–50°F (4.5°C–10°C) mark, and ideally before the first frost. Harvesting peppers not only gives you a delicious final bounty, but encourages the plant to channel its energy to preparing for winter dormancy.

TIME-SAVING TIP: Combine harvesting with an inspection for pests or diseases—multi-task to save time.

STEP 2 **Prune the Plant**
Pruning is a crucial step to redirect the plant's energy away from keeping the leaves and branches alive that will inevitably die with cold weather. Trim the plant down to a more manageable size for the winter, leaving about 6–8 inches (15–20 cm) of the main stem.

TIME-SAVING TIP: Invest in quality pruning tools for clean and efficient cuts, saving time and promoting plant health.

STEP 3 **Mulch**
Mulch with leaves, straw, or pine needles (3–4 inches [7.5–10 cm] thick) to insulate roots. In colder climates, transplant to a smaller container, mulch, and relocate to a sheltered spot near your house, in a garden shed or garage. Extreme temperatures may need a frost blanket or winter cover for additional protection.

STEP 4 **Monitor Water, Heat, and Light**
Water sparingly every few weeks just to keep the soil moist and roots alive. If growing inside, maintain temperatures of 50°F–60°F (10°C–15°C) with ambient light. Remember, the goal is to keep the peppers alive but dormant, not stimulate production.

TIME-SAVING TIP: Use a soil moisture meter for accurate watering, saving you time and preventing overwatering.

⏱ How it saves you time down the line

By overwintering peppers, you can grow your favorite and most productive peppers again, and save time. It gives you a spring head start, and allows you to kick-start the season with some plants that will burst into growth when the weather warms up, putting peppers in your harvest basket sooner. The effort in the late season pays off with earlier peppers in the spring.

Quick-Start a Fall Garden

For the time-challenged gardener, starting a fall garden in late summer is a strategic move to extend the growing season. The key is to plant cool weather vegetables that thrive in the crisp autumn temperatures that will yield steady harvests as winter approaches. Follow these guidelines, each with a time-saving tip, to quick-start your fall garden in just a few 8-minute sessions.

START FALL SEEDS INDOORS

To start your fall garden while it's still summer, the solution is simple. Start cool-weather seeds indoors under grow lights. Summer temperatures can be too intense for direct outdoor planting of these veggies. By the time cooler months roll in, you'll have healthy transplants ready for your garden, saving both time and money. This strategy allows for earlier harvests, as the plants establish strong root systems before the onset of chilly weather, allowing them to withstand light frosts and thrive into early winter.

TIME-SAVING TIP: Implement efficient indoor seed-starting methods, such as batching your seed-starting (page 94) and other tips found in the "early season" section of this chapter, to streamline fall gardening.

DETERMINE YOUR FIRST FROST DATE

Determine your first frost date of the season using a frost date calculator by entering your zip code. Once you have your frost date, count backward 8–10 weeks; this marks the ideal time to start your fall seeds indoors. Don't miss this window—planting in this time frame means you'll have flourishing seedlings to plant in your garden when the temperatures cool down in the early fall. For frost-free climate with mild winters, winter is often the prime time for cool weather veggies. Start seeds indoors for fall veggies 8–10 weeks before the temperature in your area is consistently below 75°F (24°C), typically in November or December.

TIME-SAVING TIP: Use an online frost date calculator for quick and accurate frost date predictions. Adjust planting schedules based on your specific climate to optimize growing conditions.

CHOOSE THE RIGHT VEGETABLES

Selecting the right vegetables for your fall garden is crucial for your garden to thrive. Chose cool-weather veggies that thrive in cooler temperatures under 75°F (24°C). Popular choices are greens like lettuce, mustard greens, arugula, kale, chard, and collards, as well as root veggies like radishes, carrots, and beets. Broccoli, cabbage, cauliflower, peas, and kohlrabi are also tasty additions.

TIME-SAVING TIP: Choose compact varieties that are not only space efficient, but are easier to maintain and quicker to harvest.

TRANSPLANT TIMING

Transplant your seedlings into the garden when temperatures are under 75°F (24°C), ideally 4–6 weeks before your first frost. This timeline gives the root systems time to establish themselves, allowing them to survive light frosts and continue growing and producing into early winter.

 ## How it saves you time down the line

Strategically investing short bursts of time now in late summer to quick-start your fall garden pays off with more fall veggies to harvest sooner. A thriving fall garden and steady harvests make for a delicious veggie-filled transition into winter.

Rapid Plant Rescue

As late summer arrives, our once-tidy garden beds may start resembling a jungle, with sprawling, unruly plants. An 8-minute rapid rescue mission is all it takes to transform the chaos into order, keeping plants healthy and setting the stage for late-summer harvests.

ASSESS OVERGROWN AREAS (1 MINUTE)

Take a moment to assess the overgrown areas. Identify tangled branches, sprawling vines, and any plants in need of support. Choose one plant to rescue per quick session.

GATHER SUPPLIES (1 MINUTE)

In the next minute, gather essential supplies like garden tie-up tape, twine, pruners, stakes, and shade cloth. Keeping everything within easy reach streamlines the process.

PRUNE (2 MINUTES)

Spend 2 minutes pruning excessive growth, dead leaves, or crowded branches. This manages size and helps with air circulation and sunlight penetration.

TIE AND SUPPORT (2 MINUTES)

For the next 2 minutes, gently tie up wayward vines by using stretch tie-up tape tied loosely and secured to avoid damaging plants. For taller or falling-over plants, secure them to stakes for additional support.

STRATEGIC SHADE CLOTH (1 MINUTE)

Encircle the overgrown plant in shade cloth to rein in the sprawling branches. This temporarily keeps them contained and allows you to position a larger cage for added support. After installing the bigger structure, remove the shade cloth so the plant can lean into the new support.

ROPE BRACE (1 MINUTE)

Use a strong piece of garden rope to secure a plant to a nearby structure such as a deck railing, fence, or tree. This not only provides immediate support but also encourages a more upright growth pattern, minimizing the need for future plant rescues.

⏱ How it saves you time down the line

This rapid rescue mission saves time in the long run. By promptly managing overgrown plants, you create a neater, visually appealing garden. A well-tamed garden requires less maintenance, provides easier access for harvesting, uses space effectively, and grows more veggies for a thriving late-summer garden!

Proactive Burnout Prevention

Let's keep it real: As summer winds down, many of us contemplate throwing in the gardening trowel. Scorching heat, pest invasions, and disease can overwhelm us and leave us struggling with motivation for the basic tasks.

If this sounds familiar, you might be struggling with garden burnout. The good news is you're not alone—I feel it every August! Here's a weekly routine, with each segment taking just 8 minutes. At the end of the month, you'll have a powerful routine to proactively beat garden burnout.

ACCEPTANCE AND EXPECTATIONS (WEEK 1)

- Manage expectations—challenges are part of the game.
- Pest, critters, and disease aren't your fault.
- Losses of plants and harvests are inevitable—it happens to all gardeners.
- Embrace the changing seasons; plant cycles come and go.

TIME-SAVING TIP: Do a daily quick mental reset—acknowledge challenges, accept what you can't change, then move forward.

SEEK SUPPORT (WEEK 2)

- Involve others in garden tasks—kids, partners, friends.
- Make it a shared experience—teach, spend time together, share the harvests.
- Do a little each day—it adds up to a lot by week's end!

TIME-SAVING TIP: Delegate specific tasks for each person so nothing feels overwhelming.

KEEP IT SIMPLE—STICK TO THE BASICS (WEEK 3)

- **Stick to the basics:** Watering, pruning, and harvesting.
- **Downsize if needed:** Consolidate containers and thin out garden beds.
- **Switch up your routine:** Find excitement in new, manageable tasks.

- **Refresh your space:** Clean out the old, pest-ridden, or disease-infested plants.

TIME-SAVING TIP: Focus on one area of the garden each day for daily maintenance. Commit to savoring the moment and relaxing while completing the daily task.

FOCUS ON ENJOYMENT—NOT JUST TASKS (WEEK 4)

- Take short breaks to recharge.
- Enjoy morning coffee walks—notice nature.
- Sit and relish warm summer evenings.
- Remember, it's not a race or a competition. Enjoy the journey!

TIME-SAVING TIP: Set a timer and dedicate 8 minutes to mental health and mindfulness in your garden routine.

 ## How it saves you time down the line

Feeling weary in the late summer is normal, but with just 8 minutes a week, you can keep gardening enjoyable and stay committed for the long haul. By proactively preventing burnout, you not only save the usual downtime it takes to recover, but also preserve more time and energy for gardening and for other activities you enjoy.

Time-Efficient Garden Planning

With strategic steps and time-saving tips, you can efficiently plan a thriving vegetable garden that produces delicious food and saves your precious time. Whether you're a beginner or seasoned pro, these time-saving tips will make this process a breeze.

Don't forget to grab your garden jotting journal (page 20), as writing down or drawing a quick diagram for each step will set the groundwork for an organized and efficient gardening experience. And when you document your plan, you're more likely to follow through!

Utilizing downtime in the winter dormant season avoids the spring planning rush when there's so much to do. Efficient planning leads to efficient gardening and a more abundant harvest.

STEP 1 **Strategic Proximity to Your Home**
Make your vegetable garden an extension of your home by planting a portion of it close to your house. This not only provides a pleasant view, but also simplifies harvesting, watering, and maintenance, and allows you to identify issues before they get out of control.

TIME-SAVING TIP: Keep frequently used tools nearby to minimize back and forth trips.

STEP 2 **Optimal Sun Exposure**
Sunlight is crucial for a thriving garden. Aim for at least 6–8 hours of sunlight a day for a thriving garden; more is even better. Strategically place garden beds from north to south for maximum sun exposure. Taller plants should be placed near the back to avoid shading of shorter veggies.

TIME-SAVING TIP: Observe the shadows throughout the day to identify sunny spots. Make notes or draw a diagram in your garden journal for positioning your garden beds in the upcoming spring.

STEP 3 **Proximity to a Water Source**
Situate your garden near a water source to streamline watering tasks. This eliminates the need for hauling watering cans and allows for easy installation of drip irrigation. A well-hydrated garden is a productive garden, so your veggies thrive with minimal effort.

TIME-SAVING TIP: Invest in a timer for automated watering, freeing up your time.

STEP 4 **Choose What You're Going to Grow In**
Decide what you want to grow in—raised beds, containers, in-ground garden, or a combination. Tailor your choice to fit your budget, family, and available space.

TIME-SAVING TIP: Use fabric raised beds or containers for quick setup, accessibility, reducing bending, and minimizing back strain.

STEP 5 **Plant What You Love**
Create a list of vegetables you and your family enjoy. Plan for a few new varieties each season, but prioritize space for familiar favorites so nothing goes to waste.

TIME-SAVING TIP: Check the temperature preferences of each vegetable and plant them to align with the season. Make note of this in your journal so you don't waste time growing vegetables in the wrong season.

⏱ How it saves you time down the line

By investing time in thoughtful planning during the winter, you save yourself from the chaos of last-minute decisions in the spring. This translates into less stress and more time for enjoying your garden and other activities.

Indoor Garden Once-Over

As the outdoor garden takes a winter's nap, your indoor garden takes center stage. Spend 8 minutes each week to do a thorough once-over. Not only does this keep your indoor garden thriving but it nurtures your plants to bring joy, satisfaction, and a dose of green therapy to combat the winter blues.

Sticking to the recommended time frames for each step keeps you focused on the task at hand and makes this maintenance routine a manageable and time-efficient task.

STEP 1 Prune and Pest Check (2 minutes)

Begin with a pruning session. Pinch off dry and yellowing leaves and thin overcrowded seedlings to maintain a neat appearance while redirecting the plant's energy toward healthier growth. Simultaneously, conduct a thorough pest check, examining leaves, stems, and soil for any signs of unwanted visitors. Multi-tasking pruning and pest checks saves time.

TIME-SAVING TIP: Keep essential supplies nearby in a basket including precision pruners, a small garden fork, yellow sticky traps, and a spray bottle.

STEP 2 Pest Prevention and Control (2 minutes)

Replace or add new yellow sticky traps, strategically placing them where fungus gnats are spotted. Refill DIY fungus gnat bottle caps with apple cider vinegar and a drop of dish soap (see page 12). For infestations, isolate affected plants, and weather permitting, take outside for a through spray-down to remove pests. Alternatively, do this in the kitchen sink if outdoor conditions are unfavorable. Isolate infested plants to prevent spread to other plants and spray any infestations with organic pest control.

TIME-SAVING TIP: Don't skip this step, because early detection allows for prompt intervention, minimizing time needed for pest control.

STEP 3 Soil Care (1 minute)

Use a small fork to gently loosen the top layer of compacted soil in each indoor container, working your way around the base of plants. This helps with for water absorption, aeration, and preventing waterlogged roots.

TIME-SAVING TIP: Streamline the process by batching soil care with plants in close proximity.

STEP 4 Watering and Fertilizing (2 minutes)

Water thirsty plants, making sure that soil is neither bone-dry or waterlogged (see page 16). Combine watering and fertilizing to streamline the process. Include a nutrient boost of liquid fertilizer in your watering can weekly so your plants have the essential nutrients to thrive during the winter months.

TIME-SAVING TIP: Use a watering can with a pointed spout for easy reach into containers, reducing spillage and cleanup. Place drip trays under plant containers to eliminate leaks.

STEP 5 Clean and Wipe Down (1 minute)

Cleanliness is the key to a thriving indoor garden. Wipe away dust or spilled soil from your grow lights to maximize efficiency. Periodically rinse seedling trays to remove soil and plant debris, leaving no hiding spots for fungus gnats. A clean area deters unwanted visitors and keeps your space visually appealing.

TIME-SAVING TIP: Keep a toothbrush handy for nooks and crannies. A quick swipe removes debris without disturbing your plants.

 ## How it saves you time down the line

By investing 8 minutes weekly to an indoor garden once-over, you establish a routine that not only nurtures your plants, but also prevents significant problems down the line, saving you time fixing bigger problems later. This keeps your indoor garden vibrant, healthy, and thriving throughout the dormant winter season, allowing your indoor paradise to thrive with minimal effort.

Garden Shed SOS

As the garden enters dormancy, seizing the opportunity to declutter and organize your garden shed sets the stage for a seamless transition into the busy spring garden season. Several short, focused 8-minute bursts dedicated to Garden Shed SOS will make a significant impact, transforming chaos into a well-organized space that saves time.

STEP 1 **Assess and Declutter**
Clear out your shed completely in your first 8-minute session. As you clear out, assess and declutter. Be ruthless—if you haven't used it in the past garden season, chances are you don't need it. Use the "three-box method"—keep, donate/sell, discard—to make decluttering decisions efficiently. Discard/recycle items promptly and keep other boxes handy for the next 8-minute session.

TIME-SAVING TIP: Keep trash bags, donation boxes, and recycling bins nearby for quick disposal during the decluttering process. This minimizes interruptions and keeps the momentum going.

STEP 2 **Categorize and Sort**
Categorize remaining items in your next 8-minute session. Group similar items together, such as tools, containers, fertilizers, and pest- and disease-control products. Use clear storage bins to keep everything organized and accessible.

TIME-SAVING TIP: Prioritize speed over perfection to quickly sort items. Fine-tuning can be done later, but the goal for now is to quickly group similar items together in broad categories to streamline the process.

STEP 3 **Maximize Vertical Space**
In your next 8-minute session, install shelves, hooks, and pegboards on the walls to store tools and equipment and to maximize vertical storage. Spend 8 minutes installing each storage solution, focusing on one area at a time. Use this time to hang long-handled tools vertically and organize smaller items on shelves.

TIME-SAVING TIP: Plan your vertical storage layout ahead of time to optimize space and save time during the session.

STEP 4 **Create Zones**
Designate specific areas of your shed for various functions such as potting and planting, seed starting, fertilizers, and tool storage. Keep frequently used items within arm's reach to minimize time spent searching for them.

TIME-SAVING TIP: Use colored tape or labels for each zone to help you quickly locate items and maintain organization during busy gardening seasons.

STEP 5 **Implement Time-Saving Techniques**
Schedule regular 8-minute Garden Shed SOS sessions to maintain organization. Use inexpensive storage solutions such as repurposed containers, stacking bins, clear shoe bins, and hanging baskets to keep small items organized.

TIME-SAVING TIP: Set a recurring reminder or calendar appointment on your phone to prompt Garden Shed SOS sessions. Consistency is key to maintaining organization and preventing clutter buildup.

 How it saves you time down the line

By investing several 8-minute Garden Shed SOS sessions, you'll transform from cluttered chaos into an efficient space that helps you work smarter, not harder. The time-saving techniques streamline the organizational process, allowing you to maximize limited time and energy. With an organized shed, you'll have more time to focus on the joy of gardening.

Swift Onion Seed Starting

As the garden settles into dormancy, it's the perfect time to get a head start on onions, a kitchen staple in the vegetable garden and one of the earliest crops to start from seed indoors. Onions are not only quick, easy, and economical to grow from seed, but you'll also yield flavorful harvests that will store well for the long term. In just one or two swift 8-minute seed-starting sessions, your onion seeds will be off to an early start and well on their way to thriving bulb production.

STARTING ONION SEEDS INDOORS

During one of your Seed-Starting Saturday or Sunday sessions (see page 94), utilize your seed-starting station (see page 136) to start your onion seeds.

STEP 1 **Know Your Frost Date**

Onions need an early start for proper bulb for-mation, and onion transplants will thrive in seed starting containers for an extended period. You can start onion seeds as early as 3–4 months before your last frost date. Use an online frost date calculator to quickly find this date for your area.

STEP 2 **Choose Seeds over Sets**

There are two primary methods for growing onions: from seed or from sets (tiny onion bulbs in early stages of growth). While sets are conve-nient, starting from seed allows for wider variety and often yields larger bulbs, making them a cost-effective choice for long-term storage.

STEP 3 **Understanding Your Onion Zone**

Understanding your onion zone is vital to select varieties well-suited to your area. Onion bulb for-mation is sensitive to day length, with long-day onions requiring 14+ hours of daylight for bulb formation, suitable for northern regions. Short-day onions thrive in southern climates with 10+ hours of daylight. Day-neutral onions are adapt-able to varying day lengths, making them ideal for regions in between.

STEP 4 **Prepare Seed-Starting Mix**

Utilizing your seed-starting station (see page 136), fill seed trays or containers with seed-starting mix, allocating one six-pack for each onion variety you plan to grow.

STEP 5 **Planting Seeds**

Plant onion seeds about ¼-inch (0.5-cm) deep, sprinkling a pinch of seeds lightly in each cell or container. Onion seedlings can be easily sepa-rated once they reach transplant size. Cover with a light layer of seed-starting mix and mist lightly to maintain moisture during germination.

STEP 6 **Provide Optimal Growing Conditions**

Place the seed trays under grow lights for 10–12 hours for healthy growth, keeping soil moist but not waterlogged. With proper care, your onion seedlings will thrive and be ready for transplanting once the last frost date approaches.

How it saves you time down the line

Starting onions in the dormant season is a smart gardening strategy that allows you to get a head start on your crop. Early planting allows for sufficient time for bulb development before warmer temperatures set in. This saves both time and money and gives you a bountiful harvest. With just a little bit of planning and a few swift 8-minute seed starting sessions, you can efficiently sow your seeds and enjoy flavorful onions that store well over time.

Soil Turbocharge: Quick Cover Crops for Busy Gardeners

As the garden season transitions into dormancy, a short 8-minute window of time can revitalize your soil and lay the foundation for a thriving spring garden. Often overlooked cover crops play a vital role in soil health even during winter freezes.

COVER CROPS—SOIL HEALTH SUPERHEROES

Cover crops are fast-growing plants ranging from grains and legumes to root vegetables and greens, primarily planted in fall and winter. They are not grown for food, but to cover, protect, and enrich the soil during the off season. As they decompose, they add organic matter, boost productivity, even fix nitrogen from the air, while also suppressing weeds and preventing erosion.

PLANTING COVER CROPS—QUICK TIPS

STEP 1 Clear and Prep the Soil

For efficiency, focus on one or two garden beds at a time to plant your cover crops. Clear out old plants and loosen the soil for optimal germination.

STEP 2 Choose Cover Crop Seeds

Clover, peas, mustard greens, kale, or root vegetables such as daikon radishes are hardy choices that can withstand winter conditions.

STEP 3 Sow the Seeds

Since the goal of cover crops is to shield the soil rather than harvest vegetables, plant seeds densely for even soil coverage.

STEP 4 Water and Mulch

Keep seeds moist, especially during periods of low rainfall. Mulch to retain soil moisture and protect the seeds from cold.

STEP 5 Chop and Drop

Several weeks before spring planting, chop down the cover crops at ground level and leave the roots in the soil, where they will decompose, enriching it for the next growing season. This simple yet effective technique primes your garden soil for optimal growth.

 ## How it saves you time down the line

In the few minutes it takes to plant cover crops, you can reap significant benefits in the long run. By nourishing the soil, controlling weeds, managing erosion, and improving soil health, cover crops pave the way for more bountiful harvests come spring. Invest a bit of time now, and watch your garden thrive with minimal effort.

10 Minutes

CHAPTER FOUR

Quick Wins in 10 Minutes

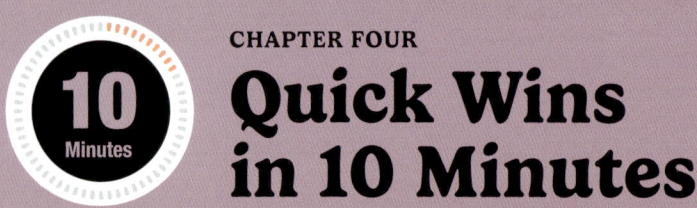

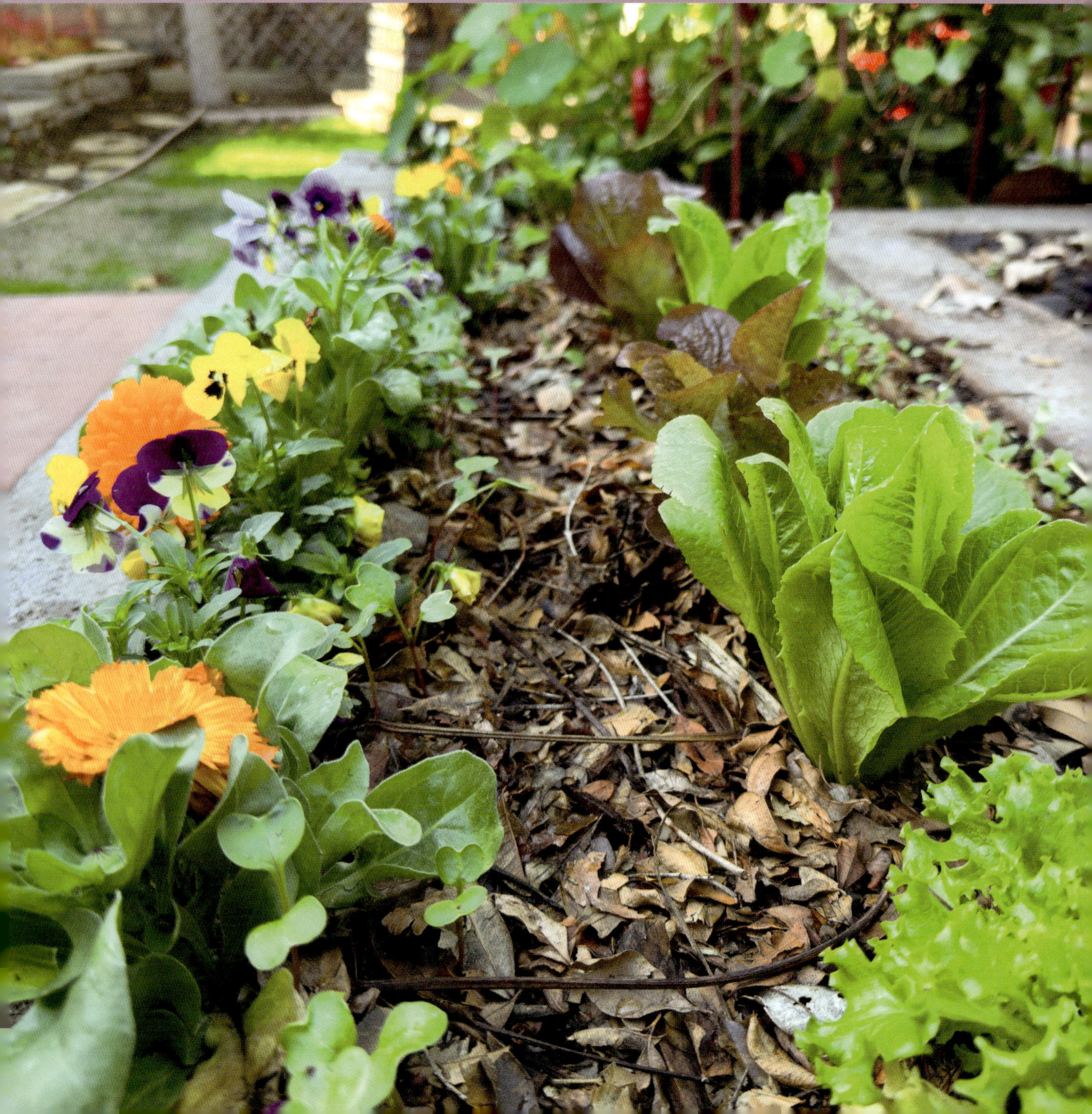

Smart and Simple

Fitting 10-minute garden tasks into your day is key to achieving quick wins that accumulate into significant accomplishments by week's end. Unlike shorter 3- and 5-minute tasks, 10-minute sessions offer a sweet spot of time that allows for more substantial progress while still being manageable within a busy schedule. Here's how to seamlessly integrate them into your day without skipping a beat.

MORNING MOMENTUM
Kickstart your day with a 10-minute garden session before diving into your daily routine. This early burst of productivity sets a positive tone and lays the groundwork for accomplishment during the day.

MID-DAY BREAK
Use your lunch break to tackle a 10-minute garden task. Whether it's filling a few containers with soil, pruning a tomato plant, watering your flowers, or soil amending, this mid-day pick-me-up refreshes your mind and energizes your afternoon.

EVENING UNWIND
Wind down at the end of the day with a calming 10-minute garden session. This helps you decompress and transition into relaxation mode.

WEEKEND BOOST
Dedicate a few 10-minute sessions throughout the weekend to tackle larger multi-step projects. Breaking these tasks into manageable chunks means you'll make progress without overwhelming your schedule.

By incorporating these 10-minute garden tasks into your daily routine, you'll enjoy quick wins and make steady progress in your garden by the end of the week. These brief but impactful sessions allow you to make the most of your time while nurturing your green oasis with care and efficiency.

Seed-Starting Station

The key to kick starting your spring garden is to create a dedicated, orderly seed-starting station stocked with all the essential supplies. With everything pre-prepped, you'll be ready to jump into a quick seed-starting session at a moment's notice and will be amazed at how many seeds you can start in a short time frame It's a guaranteed quick win! (Also see page 94.)

Create your station in just 10 minutes, gathering these supplies:

▶ **Pre-Made Seed-Starting Mix:** This mix is a game-changer. It's lightweight, sterile, and helps seeds germinate quickly. Preparing it and storing it in a bin saves valuable time every time you're ready to start seeds. My favorite recipe is a blend of 3 parts rehydrated coco coir, 1 part vermiculite, and 1 part worm castings. Mix all ingredients together in a 24-quart (23-l) storage tub, let it dry out so it doesn't mold, then put a lid on the bin and moisten as needed.

▶ **Plant Tags and Markers:** With plant tags and markers handy, you won't struggle to remember which tomato varieties you planted or mix up your cucumber and your zucchini seedlings. No more guesswork—just organization put to work for you.

▶ **Seed-Starting Trays and Six-Pack Seed Cells:** These 10 x 20 inch (25.5- x 50-cm) trays keep your planted-up six-packs organized and catch the drips. Uniform-sized six-packs are compact, stackable for easy storage, use less soil compared to larger containers, and are perfect for starting multiple seeds in one quick session, making the process efficient.

▶ **Watering Can or Spray Bottle:** A must for pre-germination watering, allowing you to water gently without disturbing soil or seeds.

▶ **Plastic Mat:** A small, reusable foldable plastic mat to place under your seed starting area keeps it mess free and simplifies cleanup. My favorite is one that snaps together at the corners to minimize the mess.

▶ **Marker:** A trusty, permanent marker makes for clear labeling.

 ## How it saves you time down the line

Having supplies pre-prepped and well-organized in one place streamlines the seed-starting process, minimizing setup and cleanup time. Clear labeling prevents confusion and helps you care for seedling-specific needs effectively and efficiently. The payoff? Robust, flourishing seedlings that yield more veggies quicker and accelerate your harvest.

Super-Speedy Grow Light Setup

You've started your seeds indoors, now what? Often a sunny window is not enough for optimal seedling growth. They need strong, intense, direct light overhead for healthy development. This super-speedy clamp light makes grow light setup a quick 10-minute task, saving you time, and makes sure your seedlings thrive, setting them up for success in the garden.

CLAMP LIGHT SETUP

The simplest and most budget-friendly option, the clamp light setup concentrates light directly onto your seedlings for strong growth. With no assembly required, it's prefect for a small space or a small tray of seedlings, making it a hassle-free solution.

MATERIALS YOU'LL NEED

▶ Clamp light with an 8.5-inch (22-cm) or 10-inch (25.5-cm) cone-shaped metal reflector

▶ CFL or LED screw-in light bulb with specified lumens (1500–2000) and kelvin (5000–6000)

▶ Clamp surface, such as shelf or cabinet

▶ A shoebox or small blocks of wood to adjust the height of the seedling tray

HOW TO MAKE A CLAMP LIGHT SETUP

STEP 1 Screw the light bulb into the clamp light fixture.

STEP 2 Clamp the fixture onto a shelf or cabinet, plug in, and turn on the light.

STEP 3 Position the seedling tray directly beneath the light, no more than 2–3 inches (5–7.5 cm) away, using a shoebox or blocks of wood to adjust the height and prevent leggy seedlings.

STEP 4 Adjust the height of the seedling tray as the plants grow, maintaining optimal distance from the light.

STEP 5 Leave lights on for 24 hours a day until seedlings germinate for intense light exposure from the moment of germination. Once the seeds germinate, turn lights on for 18 hours a day, off for 6 hours.

TIME-SAVING TIP: Set up a timer for your grow light for consistent light exposure and to avoid forgetting to turn lights on and off.

DIY CLAMP LIGHT STATION

If you don't have a shelf or cabinet to clip the light onto, creating a DIY clamp station is easy and affordable.

MATERIALS YOU'LL NEED

▶ 1-gallon (4-l) jug

▶ Sand

▶ 12-inch (30.5-cm) length of PVC pipe

STEP 1 Fill a gallon (4-l) jug halfway with sand and insert the PVC pipe.

STEP 2 Attach the clamp light to the PVC pipe.

STEP 3 Place seedling tray directly underneath the clamp light, adjusting the height as needed.

TIME-SAVING TIP: Use a funnel to fill the gallon jug, avoiding mess and saving time in cleanup.

 ## How it saves you time down the line

A 10-minute investment in setting up optimal lighting for your seedlings prevents leggy growth, saving you time and effort restarting seedlings. This quick win results in robust, healthy plants ready for transplanting in the garden, bringing fresh veggies to your plate sooner.

Hustle and Bustle Garden Bed Prep

Preparing your garden beds is an exciting part of the gardening journey. With a little hustle and bustle, you can smoothly transition from seedlings to flourishing plants. Tackle this in the early season, so you're ready for transplanting or as soon as the soil thaws in early spring. Dive into the hustle with three simple steps, completing your garden bed prep in a few brief 10-minute sessions, depending on your garden bed's size.

STEP 1 Clear the Bed

Start by clearing any debris or weeds from your garden beds. Pull out any remaining old plants from the previous season and rake the soil to loosen it up. This step gives your seedlings a clean slate, prevents competition for nutrients, and improves soil aeration.

TIME-SAVING TIP: Use a rake with a long handle for easier maneuverability or a garden claw to easily break up compacted soil, making it easier for new plants to establish roots.

STEP 2 Amend and Water the Soil

Next, it's time to amend the soil with compost, worm castings, and organic fertilizer to replenish nutrients and improve soil structure. Spread a layer of compost or fertilizer evenly over the soil surface and use a rake to work it into the top few inches (5–7 cm) of the soil. This will provide your plants with the essential nutrients they need to thrive and promote healthy root growth. If the soil is dry, water and work the moisture into the soil until it's evenly moist.

TIME-SAVING TIP: Use a soaker hose or drip irrigation system (see page 180) to water garden beds efficiently, saving time and water compared to traditional watering methods.

STEP 3 Mulch and Water

Finish off by applying a layer of mulch (shredded leaves or shredded straw) to help retain moisture, suppress weeds, and regulate soil temperature. Spread a 2–3-inch (5–7.5-cm) layer of mulch evenly over the soil surface. After mulching, give your garden beds another watering to settle the mulch and help hold it in place during wind.

TIME-SAVING TIP: Keep a mulch stash in a trash can near your garden beds for quick, hassle-free access.

How it saves you time down the line

By investing a few 10-minute bursts to prepare your garden beds in advance, you'll streamline the planting process and give your plants to best possible start. Your garden beds will be primed and ready to go as soon as your seedlings are hardened off and the weather is right to plant them. You'll be rewarded with a smoother planting process and healthier plants in the long run. Embrace the hustle and bustle and watch your garden thrive!

10 Minutes

Streamlined Seedling Transplanting

Your last frost date has passed, your garden beds are prepped, and seedlings are hardened off—it's time to kick off the garden party! Transplanting your seedlings into their new home is a quick and rewarding step. With a few 10-minute sessions spread out over several weeks, your seedlings will be well on their way to thriving. Soon, your garden will be bursting with thriving, healthy plants!

Before diving into transplanting, make sure you've completed the Hustle and Bustle Garden Bed Prep (page 140). This prior preparation streamlines the process and saves time.

STEP 1 Planting Hole Prep
Dig holes twice as wide as the seedling's starter container. For tomatoes, dig twice as wide and deep, as they can be planted up to their last set of leaves for sturdier plants.

TIME-SAVING TIP: Use a garden auger attached to a cordless drill to quickly dig multiple planting holes.

STEP 2 Soil Boost
Add a handful of compost and worm castings to each planting hole to boost nutrients. Even though you added it in the garden bed prep stage, this additional nutrient boost jumpstarts plant growth. Compost and worm castings provide gentle, time-released, slow and steady nourishment without the risk of burning your plants.

TIME-SAVING TIP: Pre-mix compost and worm castings in a wheelbarrow or large container for quicker distribution.

STEP 3 Seedling Removal
Gently free your seedling from their starter container by squeezing the bottom. Carefully remove the seedling, supporting the stem between your fingers.

TIME-SAVING TIP: Use a small spoon to delicately loosen the seedling from the container, taking care not to damage the roots.

STEP 4 Fill and Feed
Gently place the seedling in the prepared hole and fill the hole with soil. Add another handful of compost for an additional nutrient boost as you fill.

TIME-SAVING TIP: Keep an extra bucket of compost within arm's reach for easy access while transplanting.

STEP 5 Hydrate
Give your freshly transplanted seedlings a deep watering with a drink of worm tea and nitrogen-rich fertilizer, thoroughly drenching the soil for root absorption.

TIME-SAVING TIP: Mix worm tea and nitrogen-rich fish fertilizer together into your watering for a combined application.

STEP 6 Label
Label your seedlings with plant markers to easily identify them after transplanting.

TIME-SAVING TIP: Pre-label plant markers with a permanent marker before your transplanting session to expedite the process.

 How it saves you time down the line

By following these streamlined steps and implementing time-saving tips, your seedlings will be comfortably settled into their new home in the garden in no time. Providing extra nutrient boosts during transplanting gives them a strong start, fostering healthy growth and robust root development. Enjoy watching your garden spring to life with vibrant growth and delicious produce!

Perpetual Salad Bar

Ready to transform your lettuce patch into a perpetual salad bar? With just 10 minutes a week, turbocharge your harvest and extend it through the early cooler spring season, saving valuable time and effort in the long run.

CUT-AND-COME-AGAIN HARVEST METHOD

Instead of pulling the entire plant, selectively cut or pinch mature outer leaves, leaving inner leaves to continue growing. This "cut-and-come-again" method encourages perpetual growth for a season of crisp, homegrown salads. Schedule a weekly 10 minutes for this method, keeping scissors handy for quick harvests and providing a week's supply of salad greens.

TIME-SAVING TIP: Store washed and dried lettuce between layers of paper towels for longer storage of quick grab-and-go salads.

SUCCESSION PLANTING FOR CONTINUAL HARVESTS

Allocate another 10 minutes every 2–3 weeks for succession planting of lettuce seeds to maintain a steady supply without overwhelming your kitchen.

TIME-SAVING TIP: Set a recurring calendar reminder on your phone to maintain consistent planting sessions.

ROTATE HARVEST AREAS

Rotate harvest areas within your lettuce containers or garden beds weekly to prevent overharvesting from one area. This allows lettuce to regrow, prolonging your harvest window while maintaining balanced growth.

TIME-SAVING TIP: Keep track of your harvest rotations in your garden jotting journal, making for effortless planning of future planting and harvesting schedules.

PLANT DIFFERENT VARIETIES

Spend 10 minutes researching and choosing lettuce varieties suited to your local climate. By selecting heat-tolerant varieties for warmer months and cold-hardy options for cooler months, you'll optimize lettuce production and minimize maintenance.

TIME-SAVING TIP: Consult with local garden centers, and talk to friends and neighbors for tried and true varieties, reducing the need for experimentation.

PREVENTATIVE MEASURES AGAINST BOLTING

Shield your plants from excessive heat with shade cloth to extend their lifespan and prevent bolting.

TIME-SAVING TIP: Clip shade cloth to garden beds or containers for quick coverage on hot days.

FEEDING FOR CONTINUED GROWTH

After each harvest, turbocharge lettuce growth with compost tea or a liquid organic fertilizer that is easily absorbed, such as worm tea. This replenishes essential nutrients, promotes new growth, and boosts resilience against pests and diseases, giving you prolonged harvests.

TIME-SAVING TIP: For quick and efficient feeding, use a hose end sprayer with adjustable settings to customize the fertilizer concentration and coverage.

⏱ How it saves you time down the line

This strategy reduces trips to the store for fresh greens. With a consistent supply of homegrown lettuce, you'll spend less on store-bought salads, and enjoy the convenience of fresh, flavorful, colorful greens from your own perpetual salad bar all season long.

Monthly Irrigation Checkup

10 Minutes

Inspecting your drip system for leaks or clogs is a quick and simple task, crucial to do mid-season after the summer heat may have damaged the hoses. Maintaining a well-functioning, leak- and clog-free drip irrigation system not only boosts garden productivity, but also saves you valuable time and money through efficient watering practices.

Here's a quick 10-minute check you can perform on a monthly basis (see page 180 for how to set up). Any repairs that are needed can often be completed in another 10-minute session, depending on the required fixes.

TIME SAVINGS

Automated Watering: A drip irrigation system on a timer allows you to set it and forget it, automating watering, for consistent targeted moisture to your plant's root zone where it needs it the most, saving you both time and money.

Targeted Watering: Drip irrigation provides your garden precise watering only where your plants need it. This avoids wasted water and reduces time spent hand watering or adjusting sprinklers and hoses.

10-MINUTE CHECK

Performing a quick check takes just 10 minutes and prevents ongoing leaks or clogs down the line, that if left unchecked can waste water and lead to inconsistent soil moisture.

STEP 1 Inspect for Excess Moisture

Walk along the irrigation lines in your garden beds and containers and look for areas with excess moisture in the soil, or puddling, indicating a potential leak.

STEP 2 Run a Manual Check

Push the manual button on your timer to run a quick check. As the water is running through the hoses and emitters, look for any spurting or puddling water, indicating a leak or break in the line.

STEP 3 Repair Leaks and Clogs

Keep a drip irrigation connector kit handy for quick repairs. This kit should include various connectors, ¼- and ½-inch (0.5- and 1-cm) hoses, goof plugs, compression fittings, pliers, a cutting tool for easy cutting, and installation of replacement hoses and fittings.

STEP 4 Timer Check

If your drip system is connected to a timer, quickly review settings and make adjustments based on temperature changes or precipitation. Change timer batteries if needed.

 ## How it saves you time down the line

By incorporating this monthly 10-minute irrigation check, you can identify and fix existing or potential issues quickly, ensuring that your drip irrigation system works consistently and effectively to save you watering time and help you grow more veggies.

Fast-Track Tomato Techniques

Tomatoes are the star of the summer garden, but with strategic mid-season planning and a few 10-minute sessions using three fast-track techniques, you can enjoy the fresh tomato flavor into the fall and even through winter in frost-free climates.

TECHNIQUE #1: TOMATO CLONING: HARNESS THE POWER OF SUCKERS

Propagate (or clone) new tomato plants quickly from healthy suckers, the shoots that emerge between the main stem and branches.

TIME-SAVING TIPS:

- Select vigorous, leafy suckers that are at least 3–4 inches (7.5–10 cm). Remove the leaves from the bottom half of the sucker.

- Prepare the cloning jar by filling it halfway with water and a few drops of worm tea, a natural rooting hormone.

- Place suckers in the jar on a windowsill out of direct sunlight, submerging the lower half of the stem in water. Change water every few days to maintain root health.

- Once roots develop (in a few weeks), place cutting into a small container of soil.

- Once roots are established in the soil container (indicated by slight resistance when gently tugged), transplant to a larger container or garden bed, hardening off before placing outdoors.

TECHNIQUE #2: START COLD-TOLERANT TOMATOES

Jumpstart fall tomato harvests by starting cold-tolerant tomato varieties from seed indoors under grow lights in the mid-season for fresh tomatoes to harvest as temperatures drop.

TIME-SAVING TIPS:

- Choose quick-maturing cold-tolerant varieties like Siletz, Stupice, and Sub-Arctic Plenty (a.k.a. Chilly Willie).

- Transition seedlings outdoors once they're 6–8 inches (15–20 cm) tall.

- Harvest partially green tomatoes to protect them from an early frost and enjoy sooner.

TECHNIQUE #3: SUCCESSION PLANT DWARF VARIETIES

Compact dwarf tomato varieties are ideal for container gardening, offering portability and flexibility to extend your harvest and grow inside during cooler months.

TIME-SAVING TIPS:

- Choose container-friendly varieties 1–3 feet (30 cm–1 m) tall, such as Tiny Tim and Golden Nugget.

- Start seeds indoors under grow lights to get them off to a quick start.

- Transplant into 5-gallon (19-l) fabric containers with handles for portability and good drainage. Move containers indoors during cooler nights or grow indoors for winter harvests.

- Maintain high production by fertilizing weekly.

How it saves you time down the line

Incorporating these techniques into your mid-season routine each year fast-tracks your late summer and fall tomato harvests, extending them into the cooler months. This smart gardening strategy allows you to grow more in less time, enjoy the amazing homegrown tomato flavor, and avoid buying sub-par tomatoes at the grocery store in the off-season.

10 Minutes

Quick-Fix Hanging Basket Refresh

Are your hanging baskets looking tired and worn out in the warm summer mid-season? Don't worry! With just four easy steps in about 10 minutes, you can quickly refresh and breathe new life into them, keep them blooming, and produce veggies again in no time.

STEP 1 **Remove Spent Flowers and Veggies (3 minutes)**

Start by removing any spent flowers, wilted foliage, or overgrown veggies from your hanging baskets. This step not only improves the appearance, but also stimulates new growth and blooming.

STEP 2 **Freshen Up (3 minutes)**

Add fresh flower or veggie transplants if space allows. Chose vibrant colors and healthy plants to inject a pop of color for a beautiful focal point that you'll see and enjoy.

STEP 3 **Add Fresh Soil and Nutrients (3 minutes)**

Once you've cleared out the old vegetation and added some new plants, it's time to replenish with fresh soil and nutrients. Loosen compacted soil and incorporate a blend of fresh potting mix, compost and worm castings, and slow-release fertilizer. Mix these amendments into the soil for optimal plant growth.

TIME-SAVING TIP: Pre-mix and pre-moisten soil and amendments to streamline the process and save valuable time.

STEP 4 **Soak and Feed (1 minute)**

Give your newly refreshed hanging baskets a thorough soak to rehydrate the soil and settle the plants into their new environment. Place the baskets in large storage totes or tubs filled with water, and add some water-soluble liquid fertilizer for a quick nutrient boost. Allow the baskets to soak for a few hours for thorough hydration and nourishment.

TIME-SAVING TIP: While your baskets are soaking, take advantage of this time to tackle other garden tasks, or simply relax and enjoy some well-deserved down time.

 ## How it saves you time down the line

By following these four simple steps, you can quickly breathe new life into your hanging baskets. With just 10 minutes of your time, you'll transform them from tired and worn out to vibrant focal points for your garden with minimal effort for maximum enjoyment.

Garden Grocery Store Grab 'n Grill

Grilling from the garden grocery store turns garden-fresh produce into delicious meals that last all week, in just 10 minutes! It's more than meal prep, it's a cherished summer tradition for us. Gathering friends for grilling sessions or harvesting together keeps the job manageable and enjoyable.

STEP 1 Garden Bounty Gathering

Head to the garden grocery for ripe, vegetables such as squash, eggplant, and colorful peppers. Snip garden-fresh herbs for added flavor.

TIME-SAVING TIP: Designate an "aisle" of the garden grocery store for each helper. Keep garden shears and various harvest baskets ready for efficient shopping.

STEP 2 Streamlined Preheat

While washing and chopping veggies, preheat the grill. This dual-tasking saves valuable time, so that everything is ready to roll when your veggies are prepped.

TIME-SAVING TIP: Use a grill basket for quick flipping or a tray with small holes so the veggies won't slip through the grill racks.

STEP 3 Punch Up the Flavor

Keep it simple but flavorful with a drizzle of olive oil, chopped garden-fresh herbs, sea salt, and freshly ground pepper.

TIME-SAVING TIP: Prepare herb-infused olive oil in advance for a quick burst of flavor.

STEP 4 Grill to Perfection

Place seasoned veggies on the grill, stir-frying, flipping for even cooking and a slight char. No need for perfect rows, this is more of a stir-fry approach.

TIME-SAVING TIP: Cut veggies into larger pieces to minimize flipping time and for even cooking.

STEP 5 Enjoy the Grilled Goodness and Batch Portion

As veggies come off the grill, have plates ready to enjoy your garden-to-table goodies. Portion the rest into meal-prep containers, letting cool completely before sealing.

TIME-SAVING TIP: Invest in quality, reusable, microwave- and dishwasher-safe containers for easy reheating and cleaning.

 How it saves you time down the line

Investing just 10 minutes up front to prep and grill veggies saves time cooking and cleaning throughout the week. Enjoying a garden-to-table meal with your loved ones brings extra satisfaction with every bite. With ready-to-go nutritious meals at your fingertips, you'll have more time for other activities or simply relaxing.

Pollinator Power Plan

10 Minutes

A strategy to attract pollinators is essential for a thriving garden. Providing a steady nectar supply draws in pollinators and beneficial insects that boost your harvest. This approach not only keeps your garden buzzing but also significantly increases the quantity and quality of your crops.

PLANT IN WAVES

Planting flowers in waves (also known as succession planting) involves strategically planting seeds or seedlings multiple times throughout the growing season. This method gives your garden continuous blooms and provides a steady food source for pollinators. By spending 10 minutes every few weeks on this strategy, you'll enjoy blooms all summer long and keep the pollinators coming in droves to your garden.

BENEFITS

▶ **Bee Buffet:** By planting flowers in mid-season to bloom in late season, you'll attract and provide bees with nectar and pollen throughout the entire growing season.

▶ **Continuous Blooms:** Enjoy colorful flowers from early spring to late fall, providing a constant source of color and happiness for you to enjoy the season-long beauty.

▶ **Boosted Vegetable Harvests:** As pollinators like bees and butterflies enjoy the buffet of flowers, they'll pollinate your vegetables more, resulting in more fresh vegetables on your dinner plate!

STEP 1 Choose Pollinator Pleasers

Select quick-growing, pollinator-friendly flowers such as zinnias, cosmos, sunflowers, and wildflower mixes. Choose a mix of colors, shapes, and heights to create a visually stunning floral display.

TIME-SAVING TIP: Choose seed mixes with a variety of pollinator-friendly varieties, different bloom times, and annual and perennial flowers. This eliminates the need to choose individual flower varieties and simplifies the planting next season.

STEP 2 Plant in Waves

Strategically plan your garden layout to leave empty spaces for successive planting waves. As one wave of flowers starts to bloom, plant seeds in the gaps to have a continuous display of blooms and pollinator pleasers.

TIME-SAVING TIP: Use biodegradable seed tape to quickly plant seeds in rows without the need for individual seeding.

STEP 3 Start Seeds Outside

Start seeds in trays or six-packs outside on a garden potting bench. This eliminates the need to set up grow lights and gives you healthy transplants to replace fading flowers later in the season.

TIME-SAVING TIP: Use biodegradable seedling pots to minimize transplant shock and save time on repotting.

STEP 4 Harvest and Replant

As the first wave of flowers fade, harvest any spent blooms to encourage new growth. As these plants reach the end of their life, transplant the seeds started outdoors into their place for season-long blooms.

TIME-SAVING TIP: Enlist the help of flower-loving friends or family members to make the process faster and more enjoyable.

 ## How it saves you time down the line

By incorporating the Pollinator Power Plan as part of your gardening routine, you'll transport your garden into a pollinator's paradise with minimal effort. Staggered plantings of quick-growing flowers provide non-stop color bursts and boost your vegetable harvests. Enjoy the blooms, reap the harvests, and watch your garden thrive without breaking a sweat.

Summer Squash Super Tricks

Feeling overwhelmed in the late summer days as the summer squash plants seem to take over the garden? Spend just 10 minutes on these tricks and your garden will look neat and tidy, your summer squash will be under control, and you'll get a few more squash to harvest as summer turns to fall.

TRICK 1: PRUNE FOR PRODUCTIVITY

Put on those gloves and grab those pruners. Spend a few minutes removing the lower branches below the first set of squash or flowers, as squash plants don't need those leaves. Squash branches are hollow, but become solid at the main stem. Cut close to the main stem to minimize disease risk and redirect energy into producing more squash. This also increases airflow, reducing the chance of powdery mildew. Especially focus on leaves that have powdery mildew to prevent its spread.

TIME-SAVING TIP: Use sharp, clean pruners for cuts. Keep a large waste can nearby to minimize cleanup time.

TRICK 2: FEED TO FLOURISH

After feeding, sprinkle a handful of compost around the base of each plant and gently work it into the soil. This provides a nutrient boost that will keep your plants flourishing. For an instant perk up, give them a shot of water-soluble fertilizer like worm casting tea.

TIME-SAVING TIP: Use a hose end fertilizer attachment that automatically mixes water-soluble fertilizer for quick application.

TRICK 3: PEST PATROL

Do a quick scan for signs of aphids or other pests. A strong spray of water can often dislodge them, while organic pesticides like neem oil might be needed for infestations. Spray in the morning or evening, not the heat of the day, so as not to burn your plants.

TIME-SAVING TIP: Always do a test spray first and keep a spray bottle handy for on-the-go pest management.

TRICK 4: HAND POLLINATION

As bee activity dwindles, take matters into your own hands—literally. Identify male and female flowers and use a cotton swab to transfer pollen between them.

TIME-SAVING TIP: Keep cotton swabs in your pocket to streamline your routine.

 How it saves you time down the line

By investing just 10 minutes in these simple tricks, you'll not only have a few more squash to harvest late in the season, but also save valuable time down the line (tell me how, please!). With well-maintained plants, you'll enjoy a beautiful thriving garden in the waning days of summer. Plus, by preventing overgrowth, you'll reduce the need for extensive maintenance, allowing you to relax, enjoy, and savor the fruits of your labor.

Quick-Plant Fall Garden

As the late season arrives, it's time to enjoy crisp, autumn days before winter. With cooler temperatures, dive into quick-starting a fall garden of cool-season veggies, putting those seedlings we started back on page 118 to work. In just a few 10-minute sessions, we can have seedlings snug in the ground and ready to thrive! Scale down to one to two raised beds or containers to give avoid burnout after a busy summer.

STEP 1 **Strategic Planning**

Kickstart your fall garden by mapping out a planting schedule in your first 10-minute session. Review crop maturity times on seed packets and count backward from the first frost date in your area to determine the optimal planting time for each. Add a few weeks to account for slower growth due to shorter days and cooler temperatures. By scheduling carefully, you'll fit harvests in before winter.

TIME-SAVING TIP: Use your garden jotting journal to record planting dates, streamlining future fall gardens.

STEP 2 **Preparing Raised Beds/Containers**

In your next 10-minute quick-plant session, prep your planting space by clearing out spent summer crops. Amend with compost, worm castings, and granular fertilizer.

TIME-SAVING TIP: Consider installing drip irrigation (see page 180) to save time and water.

STEP 3 **Transplanting Seedlings**

With prep work done, transplanting seedlings is a breeze in another 10-minute session. Water seedlings before transplanting to make them easier to remove from containers. Dig holes slightly larger than the seedling containers and place each seedling in its planting hole. Water thoroughly with worm casting tea mixed in to minimize transplant shock.

TIME-SAVING TIP: Mulch with shredded leaves to protect them from late-summer heat waves or unseasonably cold nights.

STEP 4 **Prepare Shade Cloth/Cold-Season Covers**

Late summer/early fall temperatures can be unpredictable. Spend your final session preparing shade cloth and cold season covers. Keep a selection of covers ready for use, such as shade cloth, row covers, or cloches to protect vulnerable seedlings. Monitor forecasts and be ready to deploy covers as needed to protect plants from heat waves or cold damage.

TIME-SAVING TIP: Pre-cut covers to size and label them for quick installation during sudden weather changes.

⏱ How it saves you time down the line

By following efficient fall gardening and strategic planning, you'll maximize your harvest potential while saving valuable time. With just a few focused 10-minute sessions, your fall garden will thrive and will reward you with a bountiful fall harvest as long as possible during the cooler months.

Fast-Track Strawberry Winterizing

Strawberries are amazing perennials that will produce for 3–4 seasons and can withstand temperatures down to about 20°F (-6°C) with a little help. As strawberry harvests wind down in the late season, it's a perfect time to prep them for winter. In just 10 minutes, you can winterize your strawberries, getting them snuggled in and ready to weather the cold, so they'll come back next spring.

STEP 1 Variety Selection

Choose cold-hardy varieties like Allstar and Sparkle for cold-winter climates. My favorites, Seascape, Chandler, and Albion, thrive in mild winter climates. Selecting the right varieties means your strawberries will endure winters and thrive come spring.

TIME-SAVING TIP: Check online or ask local garden centers for recommendations on cold-hardy varieties suitable for your area.

STEP 2 Free Plant Jackpot

Keep an eye out for the runners—those little strawberry offshoots. Stick them in soil and they'll root for free plants. It's a fun and budget-friendly way to expand your strawberry patch.

TIME-SAVING TIP: Secure the roots with a bobby pin or bent wire for a quick and cost-effective solution.

STEP 3 Dormancy Trimming

In colder climates, strawberry leaves turn brown and crispy. They're not dead, just dormant! Trim back these leaves at the crown (where the leaves emerge) to conserve plant energy and promote new growth for next season.

TIME-SAVING TIP: Use sharp precision pruners for clean and speedy trimming.

STEP 4 Mulch Magic

Before the ground freezes, tuck your strawberries under a cozy blanket of mulch, 3–4 inches (7.5–10 cm) deep. This protects their roots from harsh winter temperatures, so they stay snug until spring. For container plants, move to a sheltered spot and cover with a frost blanket.

TIME-SAVING TIP: Use shredded leaf mulch for easy application in between small strawberry plants.

STEP 5 Winter Watering Wisdom

During dormancy, keep strawberries evenly moist; don't let them completely dry out. However, avoid waterlogging, which can cause roots to rot.

 ## How it saves you time down the line

With a little up-front preparation, your strawberries will thrive come spring with new growth, growing delicious berries season after season with minimal fuss. You'll save the time and money planting new berries each year and will enjoy bountiful berry harvests.

Fuss-Free Garlic Growing

Garlic is the ultimate late-season must-plant crop for maximum flavor with minimal effort. Individual cloves yield new bulbs, offering huge return with little hassle. In about 10 minutes or less, you can plant a crop of garlic and forget about it while it overwinters. By spring, it will wake up, adding tons of flavor with very little maintenance on your part.

TOP TIPS FOR FUSS-FREE GROWING

Choose the Right Variety
Northern gardeners with cold winter climates will have the best results with hard-necked garlic, while soft-necked garlic thrives in southern frost-free regions.

Grow in Containers
Containers provide mobility and optimize space, and provide loose, obstruction-free soil for good bulb development. Choose containers at least 8–10 inches (20–25.5 cm) deep for good root development.

Prep Potting Mix
Moisten the potting mix prior to adding it to the containers to streamline prep. If reusing potting mix, add in compost for nutrient rich soil.

Separate Bulbs
Before planting, separate garlic bulbs into individual cloves, leaving the paper covering on for protection. In southern climates, simulate the required winter dormancy by freezing the bulbs for a few weeks to jumpstart growth.

Planting Technique
Plant individual cloves a few weeks before first frost in northern climates and after temperatures have cooled down to below 80°F (27°C) on a consistent basis in southern climates. Plant 6–8 inches (15–20 cm) apart with the pointed side up, flat end down, 2–3 inches (5–7.5 cm) deep.

Minimal Maintenance
Once planted, garlic requires little attention. Mulch with several inches (5–7 cm) of shredded leaves and provide occasional watering during dry spells. In southern climates, protect from sudden winter heatwaves by relocating containers to shaded areas.

Growth Expectations
Northern gardeners may see a few small green shoots emerge before cold winter sets in, then garlic will take a long winter's nap until warm spring shows up, at which point it starts to grow long shoots and bulbs. It needs this period of winter dormancy to develop bulbs to their fullest potential. Southern gardeners will see green shoots emerge in a few weeks that will continue to grow until early summer.

Harvest
Wait until lower leaves turn yellow and begin to dry out in mid-summer before harvesting garlic bulbs.

Curing
Cure garlic bulbs in a warm, dry place for several weeks to enhance flavor and prolong shelf life.

Storage
Store cured bulbs in a cool, dry place for long-lasting freshness and flavor.

 ## How it saves you time down the line

Garlic growing is the ultimate time-saving rewarding endeavor for the time-savvy gardener—big rewards for minimal effort. Plant it in minutes, forget it over winter, and harvest in summer. A flavorful investment that gives back many times over. Reap the rewards of homegrown goodness in your favorite recipes come summertime.

Pre-Frost Harvest Hustle

As the days grow shorter and the fall chill sets in, it's time to prepare for a possible frost that could be the end of your tender summer crops. With a quick 10-minute Pre-Frost Harvest Hustle, you can make the most of your garden bounty before frost sets in. Grab your partner or a friend to double your efforts, especially if you have a lot to harvest.

FROST SENSITIVE VS. FROST TOLERANT

Before diving into your harvest hustle, it's essential to know the difference between frost- sensitive crops and frost-tolerant crops. Frost-sensitive crops such as tomatoes, peppers, eggplant, cucumbers, and squash will be damaged or die if exposed to frost. On the other hand, frost-tolerant vegetables such as most leafy greens, peas, broccoli, cauliflower, and root veggies can endure light frosts, and it may even enhance their flavor.

TIME-SAVING TIP: Before starting your harvest hustle, make a quick checklist of the crops you need to gather. This will help you prioritize and save time once you get outside in the garden, and make sure you don't miss any crops.

HUSTLE IT!

Keep a close eye on the weather forecast as frost approaches. When frost is predicted, start your harvest hustle by gathering the frost sensitive crops. Use garden shears to cut tomatoes (even green ones), peppers, and eggplant from the vine. Gather cucumbers, squash, beans, and basil.

TIME-SAVING TIP: Designate a specific basket for each type of vegetable to save time when sorting and washing later on. Once harvested, handle with care to preserve freshness and prevent bruising. For unripe tomatoes, bring inside, place in a paper bag or between layers of newspaper in a cardboard box to ripen, or enjoy in your favorite green tomato recipes. Tomatoes and peppers can be washed and tossed whole into a freezer bag for quick preservation to pull out later in winter for sauces and recipes.

For cucumbers and summer squash, wrap in paper towels or in garden surplus saver bags (page 70) or blanch and flash-freeze squash, beans, and eggplant to use later in stews, soups, and sauces.

Certain vegetables, like root crops and winter squash, can be stored in a cool, dark, dry place for extended periods. Make sure there is plenty of air circulation to prevent spoilage.

Share your end-of-season bounty with friends, family, and neighbors, or donate extra to local food banks or community organizations.

FROST PROTECTION

If your frost-tolerant crops are still young seedlings or have recently been transplanted, cover them with frost blankets or cold covers for frost protection.

How it saves you time down the line

By investing just 10 minutes in a pre-frost harvest hustle, you can bring in the bounty, minimizing waste, and enjoy a delicious array of homegrown fruits and veggies at the end of the season. Throughout the winter months, have the satisfaction of sharing your abundance with family and friends.

Snappy Indoor Herb Garden

As the outdoor gardening season winds down, it's time to get your snappy indoor herb garden going. Herbs are easy to grow and grow well on a sunny window. With minimal effort and a few clever strategies, you can enjoy fresh herbs year round, adding huge flavor to your winter dishes and brightening up the long winter days.

INDOOR HERB GARDEN TIPS

Propagate Herbs for Indoor Growing: Rooting stem cuttings of basil or mint or dividing clumps of chives allows you to multiply your plants for free. Refer to page 80 for propagation tips.

Plant from Seed in Small Containers: Starting herbs from seed is an affordable and rewarding way to grow your indoor garden. Choose small containers 1 gallon (4 l) or less with drainage, and fill them with pre-moistened organic potting mix. Sow seeds according to package directions, keeping soil moist. Herbs such as basil, parsley, dill, cilantro, and chives are easy to grow indoors and thrive in small pots on a sunny windowsill.

Grow in Water: Some herbs such as basil and mint are very easy to propagate and grow in water for an extended period. Change the water every few days to prevent stagnation and clip as needed.

Snip and Clip to Keep Growing: Snip off the top few inches (5–7 cm) regularly to encourage branching and prevent leggy stems for fuller plants. You'll enjoy the fresh herbs in your recipes.

Plant New Seeds Every Few Weeks: Have a continuous harvest of fresh herbs through the winter by staggering your plantings. Start new seeds or transplant seedlings every few weeks to replenish your herb garden.

TIME-SAVING TIPS:

▶ Invest in self-watering small containers to maintain consistent moisture levels.

▶ Use grow lights on a timer for vibrant growth, especially during the darker winter months.

 ## How it saves you time down the line

By incorporating these quick herb tips into your indoor gardening routine in the dormant season, you can enjoy a thriving herb garden with little effort. Not only will you have access to fresh, flavorful herbs for winter cooking but you'll also experience the joy of greenery during the winter months.

Rapid Garden Research

As the outdoor garden heads into dormancy, it's tempting to give your green thumb a break. However, investing 10 minutes a few times a week during the dormant season pays dividends when the busy garden season rolls around. This period of relative quiet is the perfect time to read and research new varieties, explore new planting techniques, brush up on gardening skills, or even take an online gardening class.

WHY THE DORMANT SEASON MATTERS

The dormant season may seem like a time of inactivity, but it's actually a crucial phase in the gardening cycle. Not only does it give the garden time to rest, but it also provides you with a break to chill and focus on other activities to further your gardening skills. By investing time and effort into research now, you set yourself up for success when the new growing season begins. Whether you're researching new varieties, planning your garden layout, or learning new planting techniques, the knowledge and skills you acquire now will pay off in the months to come.

QUICK RESEARCH TIPS

▶ **Set Aside Regular Time:** Dedicate a specific time each week to research and learning, such as 10 minutes before bedtime a few times a week or on weekends.

▶ **Use Online Resources:** Watch YouTube videos, peruse gardening websites and forums, and join online community groups to connect with other gardeners. This helps with "cabin fever" during cold winter months and avoids a feeling of isolation.

▶ **Take Notes:** Jot down insights, new varieties you want to grow, what you're learning, and topics you want to dig into deeper.

 ## How it saves you time down the line

Dedicating just 10 minutes a few times a week to gardening research and learning during the dormant season saves you countless hours of trial and error during the busy gardening season. The insights you gain now will make your gardening more enjoyable in the long run. So don't overlook the dormant season—embrace it as a time to invest in your gardening success.

Prune Dormant Berry Bushes

10 Minutes

Pruning blackberries or raspberries is not just a gardening chore, but is essential to a bountiful delicious berry harvest next summer. As the garden winds down for winter, prep your berry bushes for the next growing season. With some know-how and time-saving tips, you'll have a thriving berry patch that will bring you joy all summer long in just a few 10-minute sessions, depending on the size of your berry patch.

STEP 1 Understand Canes

Blackberries and raspberries have two types of canes, or branches: the floricanes and primocanes. Floricanes produce berries during the growing season while primocanes are vegetative canes that will bear fruit the following year. Identifying and recognizing these differences will guide your pruning.

STEP 2 Gather Your Gear

Wear a good pair of gloves to protect against thorns and use heavy-duty pruners for cutting through thicker branches. These will make the pruning process quicker and safer.

STEP 3 Identify and Remove Floricanes

Locate the floricanes by their lighter brown color and remnants of fruit. They might also appear brown and dead, depending on how late into the season you prune. Follow each cane to the base and cut it off, as it will not produce again. Removing these dead canes allows the plant to redirect its energy into producing new growth and fruit for next season. Remember it's done producing, so give it a farewell snip!

STEP 4 Trim Primocanes

Next, prune the primocanes, which will become fruit bearing canes next year. Trim them to about 2 feet (61 cm) in height to promote healthy growth next spring. Keep about four to six primocanes per plant for optimal fruit production. Think of it as giving your future berries a little haircut to keep them growing and strong.

STEP 5 Work Methodically

Systematically prune each plant in your berry patch, working from left to right or any other organized pattern. This methodical approach makes sure each plant is pruned properly and the entire patch is pruned efficiently.

STEP 6 Clean Up

As you prune, keep a waste bin nearby to dispose of branches, making cleanup a snap. Plus, you'll have a tidier garden to admire once you're done, a bonus reward for your efforts!

How it saves you time down the line

By investing time pruning during the dormant season, you're setting yourself up for a successful berry harvest the following season. By dedicating short, focused sessions to the task, you avoid being overwhelmed. Just imagine the joy of picking ripe, juicy berries from your own garden on a warm summer day, the ultimate reward during the summer for your hard work in the dormant season.

CHAPTER FIVE

Beyond the Basics: 30-Minute Level-Up Bonuses

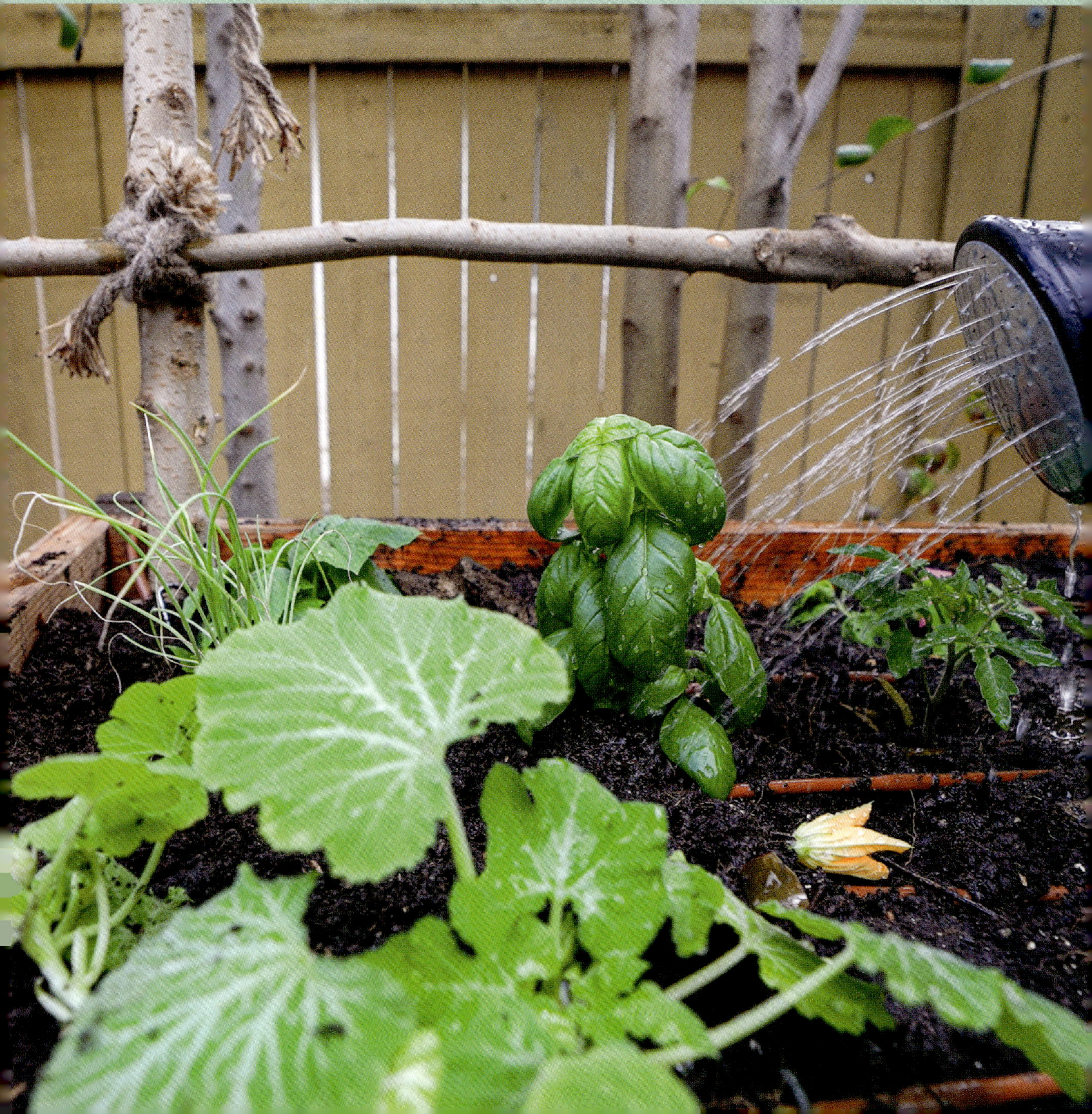

Now that you've experienced the time-saving benefits of efficient gardening, it's clear that you *can* have a flourishing garden without sacrificing your precious time. It's time to go beyond the basics to level up your garden game with 30-minute multi-step projects that will take your garden from ordinary to extraordinary.

These 30-minute projects will enhance your garden layout, improve soil health, and boost plant growth. From creating custom plant supports to installing drip irrigation, these tasks build on your foundational skills and elevate your garden's productivity and beauty.

WHEN TO FIT THEM IN

Weekends: Make the most of longer stretches of time to tackle more extensive projects like installing raised beds or building trellises. These projects require multiple steps and can significantly enhance the functionality and aesthetics of your garden.

Evenings: Use your evenings to complete a few steps of one 30-minute project after work or before dinner, gradually working toward completion.

Day Off or Mid-Week Mornings: Take advantage of a day off work or a quiet morning during the week to focus on detailed tasks that require concentration.

Enlist Help: Gardening is always more enjoyable and efficient with a helping hand. Invite friends or family to join in on your gardening adventures. Not only does teamwork make the work go quicker, but it also creates memorable experiences and strengthens relationships.

Let's level it up! Embrace these multi-step 30-minute projects, go the extra mile, and watch as your garden transforms into a vibrant and thriving paradise that you'll spend many hours enjoying.

30-Minute Composting for Busy Gardeners

Do you feel like composting is a complicated, space- and time-consuming garden task? Not anymore! In about 30 minutes with five simple steps, and just 2 feet (61 cm) of space, you can level up your garden by unlocking the magic of compost. The result? Free fertilizer to supercharge your veggies.

Compost, often dubbed "black gold," is simply decomposed organic matter. It provides important nutrients for plant growth. It enhances soil aeration and drainage, and is an eco-friendly way to turn trash into treasure that reduces your carbon footprint.

COMPOSTING IN 5 EASY STEPS

STEP 1 **Collect Brown and Green Materials**
BROWNS: Carbon-rich materials like leaves, straw, shredded plain-colored newspaper and cardboard, dry plant clippings, branches, sawdust, and pine needles

GREENS: Nitrogen-rich materials—plant-based food scraps, eggshells, coffee grounds, and green grass clippings

Collecting Materials Made Easy

KITCHEN SCRAPS: Keep a 1-gallon (19-l) bucket lined with a compostable bag under your kitchen for easy collection. Seal the bag when it's full, toss it in the freezer, and continue collecting scraps in your bucket.

GARDEN WASTE: Utilize a covered trash can outdoors for leaves, garden trimmings, and grass clippings.

Chop kitchen scraps and shred leaves with lawnmower as you collect them for quicker decomposition.

Over a few months, collect enough to fill a 50-gallon (190-l) outdoor compost container, aiming for a 3:1 ratio of browns to greens. A large pile with this ratio will heat up and transform into finished compost in 60–90 days, as opposed to 1–2 years for a cold compost pile.

STEP 2 **Layer Browns and Greens**
Once you have enough materials collected, it's time to layer. Thaw your frozen kitchen scraps outside in a laundry basket. Layer browns and greens in your compost container, aiming for the 3:1 ratio of browns to greens. No need to stress about perfection.

My go-to is the Smart Pot 50-gallon (190-liter) Compost Sak. This ready-made aerated fabric bag quickly unfolds, pops up, and is a breeze to fill, saving you the time of building a DIY container. An easy DIY option is a large trash can with holes drilled for aeration.

STEP 3 **Water and Mix**
Microbes, the tiny organisms that break down the organic matter, need water and air to thrive. Water as you layer, mixing with a pitchfork or shovel, until it's evenly moist, like a wrung-out sponge.

STEP 4 **Cover**
Cover your compost to retain moisture and heat while keeping the critters out. The Compost Sak's snug cover is ideal, or use a DIY option like heavy plastic or a trash can.

STEP 5 **Monitor Temperature and Turn**
Monitors temperature with a compost thermometer. If the pile is built correctly, it will heat up to 140–160°F (60–71°C) within 2–3 days. If the pile is not heating up, add more green materials, such as coffee grounds or grass clippings.

Turn the pile every 3–4 days to aerate it, keep the microbes happy, and facilitate an even breakdown. If the pile is dry, add water and mix in until evenly moist.

⏱ How it saves you time down the line

In 60–90 days, your hot compost pile will transform into beautiful black gold. Sprinkle a handful once a month around your plants and watch the magic unfold. Your garden will thrive and reward you with abundance, making your compost effort well worth it!

Quick DIY Tree Branch Trellis: Free and Functional

Transform tree branches in to a sturdy and rustic ladder trellis with this quick 30-minute mid-season project to support your climbing vegetables and beautify your space. This quick 5-step process can be customized to fit any garden space.

STEP 1 Gather Materials

Collect trimmed tree branches that are 1–2 inches (2.5–5 cm) in diameter and 8–10 feet (3 meters) tall. If you don't have tree branches, use purchased wooden poles.

TIME-SAVING TIP: Gather branches while trimming trees in your yard or neighborhood to save time and resources. Set them aside for project day.

STEP 2 Measure and Trim Branches

Measure and cut to your desired length and width for your trellis, at least 3 feet (1 m) for the vertical ladder supports, and 4–6 inches (10–15 cm) for the horizontal ladder rungs. Remove excess branches, leaves, and twigs for a clean and uniform appearance.

TIME-SAVING TIP: Use heavy-duty pruners or a hand saw to quickly trim branches to size. Keep a large waste bin nearby for quick cleanup.

STEP 3 Arrange Branches

Lay out the branches on a flat surface in a ladder shape, with the thicker ends at the bottom and the thinner ends at the top. Space the branches out evenly to create rungs for the ladder trellis.

TIME-SAVING TIP: Use a measuring tape to make sure the branches are evenly spaced and aligned before securing them together.

STEP 4 Secure with Twine

Use heavy-duty twine to securely lash the joints of the branches together, wrapping the twine tightly around each connection point. Tie knots to secure the twine in place, so the trellis is stable and sturdy.

TIME-SAVING TIP: Pre-cut lengths of twine and have them ready for quick and easy lashing of the branches.

STEP 5 Install in the Garden

Place the completed latter trellis in your garden, positioning it securely in the soil to support climbing plants such as pole beans or cucumbers. Train the plants to climb the trellis as they grow, keeping them off the ground and maximizing space.

TIME-SAVING TIP: Enlist the help of a family member or friend to assist with positioning and installing the trellis in the garden.

 ## How it saves you time down the line

Making your own tree branch trellis not only saves money, but also brings immense satisfaction. This DIY project brings tons of rustic charm to your garden while supporting your plants, increasing productivity. By repurposing trimmed branches and simplifying the process, you save time and money. Before long, you'll discover various uses for tree branches in the garden, so unleash your creativity and make gardening even more enjoyable!

Vegetable Garden Once-Over

Giving your vegetable garden a "once-over" is crucial in the late season to keep insects and diseases at bay and keep the harvests rolling in. This systematic inspection and maintenance routine makes a significant difference, preventing potential issues from spiraling out of control. By dedicating 30 minutes every 2–3 weeks to this 4-step systematic process, you can maintain a healthy garden and avoid overwhelming maintenance.

STEP 1 **Prune and Tie**

Start by pruning any overgrown, diseased, or dead foliage and tying up unruly plants. This improves airflow and sunlight penetration and reduces the risk of disease spread. Use sharp pruners and garden twine for quick pruning and support.

TIME-SAVING TIP: Keep pruning tools handy in a tool belt or bucket for quick access. Make quick decisions about which plants need attention to stay focused and efficient.

STEP 2 **Spray with Water**

Give your plants a gentle spray with water to knock down any pests and spider webs hiding among the foliage. This disrupts their breeding ground, reducing the risk of infestations and keeping your garden clean.

TIME-SAVING TIP: Use a lightweight flexible garden hose or a retractable hose reel for easier maneuverability and storage. Attach a wand or extension nozzle for better reach.

STEP 3 **Mulch Touch-Up**

Touch up any areas of mulch that may have shifted or thinned out. Mulch helps regulate soil temperature, retain moisture, and suppress weed growth, all critical in the late season. Add a fresh layer of mulch where needed to maintain consistent coverage and depth.

TIME-SAVING TIP: Use a lightweight mulch such as straw or shredded leaves that can be quickly spread by hand. Keep a stockpile nearby for easy access during your once-over sessions.

STEP 4 **Plant Boost**

Finally, give your hungry plants a nutrient boost with a dose of organic fertilizer, compost, or water-soluble liquid fertilizer. Late-season crops often need additional nutrients to keep them producing after multiple harvests in summer. Apply a handful at the base of each plant and water in thoroughly.

TIME-SAVING TIP: Pre-mix granular fertilizer and compost in a 5-gallon (19-l) bucket for easy application during your once-over. Use a hose-end sprayer that automatically mixes with water for quick liquid fertilizer application.

 ## How it saves you time down the line

Incorporating this once-over routine into your gardening schedule every 2–3 weeks in the late season will keep your garden healthy and thriving, and help you stay ahead of potential problems. Spending just 30 minutes systematically tending to your plants can prevent time-consuming maintenance issues down the line.

Autopilot Watering: Drip Irrigation Made Simple

Tired of endless hours watering by hand? Say hello to autopilot with drip irrigation! It may seem intimidating, but starting with a drip irrigation kit (available at most garden centers) makes it easy and time efficient. Install it in the dormant season, when the garden beds are empty, once the ground is thawed, to avoid overwhelm and be ready for spring garden season.

START SIMPLE, EXPAND LATER

Install drip irrigation in one garden bed at a time in a 30-minute session for each bed. With all the parts included in the kit, it takes the guesswork out and makes the process time efficient. Once you've mastered one bed, expanding to the others becomes a cinch. Each 30-minute session spent installing drip irrigation saves you countless hours of hand watering in the future. By focusing on one bed at a time, you'll streamline the process and avoid feeling overwhelmed.

BUILD CONFIDENCE WITH EACH INSTALLATION

As you gain experience with drip irrigation, you'll become more confident in your abilities. What once seemed daunting will soon feel like second nature. Before you know it, you'll wonder why you ever hesitated to start.

WHY DRIP IRRIGATION?

Drip irrigation systems deliver water directly to the roots of your plants, resulting in efficient watering and healthier plants. By installing drip irrigation, you'll save time, money, and water, all while promoting healthier plant growth.

STEP-BY-STEP GUIDE TO INSTALLING DRIP IRRIGATION

STEP 1 Choose Your Setup

Decide whether to connect your drip system directly to your hose bib. If your hose bib is not near your garden bed, connect a heavy-duty hose to the hose bib and run the hose to your garden bed; your supply line will be attached here.

STEP 2 Automate with a Timer

Install a timer to automate your watering schedule. Screw the timer onto your hose bib or heavy-duty hose and set your desired watering schedule.

STEP 3 Install the Basics

Lay out the parts in the kit so you can quickly identify them during the install. Attach the screen filter to your hose bib to prevent debris from clogging the system. Then, connect the pressure regulator for optimal water pressure.

STEP 4 Set Up the Supply Line

Connect the swivel adaptor to the pressure regulator and attach the ½-inch (1-cm) poly tubing supply line. Run the tubing to your garden bed and secure with landscape staples along the shorter end of your garden bed.

STEP 5 Add Drip Emitters

Punch holes every 6 inches (15 cm) in the supply line and attach the ¼-inch (0.5-cm) drip emitter tubing with barbed connectors. Run the emitter tubing along the length of the garden bed and secure it with landscape staples.

STEP 6 Repeat and Plug

Repeat the process every 6 inches (15 cm) along the supply line for even water distribution in the garden bed. Cut the emitter tubing to fit the length of the garden bed and seal the end with a goof plug.

 ## How it saves you time down the line

Autopilot watering with drip irrigation is the ultimate time-saving solution for busy gardeners. By starting simple, taking it one step at a time, and building your skills gradually, you'll save time and money in the long run and transform your garden into a thriving oasis with minimal effort. Say goodbye to hand watering and hello to a lush flourishing garden!

Conclusion

As we come to the close of *The 10-Minute Gardener*, I extend my heartfelt congratulations to you, my fellow gardener! You've embarked on a journey to unlock the secrets of gardening smarter, not harder, and I couldn't be prouder of your dedication!

Within these pages, you've discovered a wealth of time-saving tactics tailored to fit seamlessly into your busy life. From quick 3-minute tasks to more involved 10-minute endeavors, each small action contributes to the success of your garden.

My friend, I encourage you to keep this book close at hand and bring it out into the garden with you. Let it be your trusted companion, guiding you through each day's tasks with ease. And don't worry if it gets a little dirty along the way—after all, that's the sign of a well-loved garden tool!

The choice of which tasks to tackle is entirely yours. Whether planting, watering, harvesting, or simply admiring nature's beauty, every moment spent in your garden is a step closer to your dream oasis.

By applying the strategies within these pages, you'll maximize harvests, maintain happy and healthy plants, and most importantly, avoid burnout and being overwhelmed. Gardening should bring you joy and relaxation, not stress.

Your time is precious, and so is the balance you've achieved in your life. *The 10-Minute Gardener* is more than just a book, it's a guide to cultivating harmony between your garden and your life.

Let's embrace each moment, knowing that every minute spent in the garden is a testament to your dedication and passion for growing your own food. Here's to many fruitful seasons ahead!

Happy Gardening,

About the Author

CaliKim is your go-to guide for all things gardening. As an organic gardener based in sunny Southern California, CaliKim has a passion for helping busy individuals grow their own fresh, organic produce in simple and cost-effective ways.

Alongside her husband, Jerry, also known as CameraGuy, CaliKim co-owns CaliKim Garden and Home DIY, Inc., a family-owned business dedicated to empowering people worldwide to cultivate their own thriving gardens. With a wealth of experience managing a busy seed and garden shop, YouTube channel, and writing two other gardening books, *Organic Gardening for Everyone*, and *The First-Time Gardener: Raised Bed Gardening*, CaliKim understands the challenges of balancing a love for gardening with the demands of everyday life.

Drawing from her own journey of trial and error, CaliKim shares practical strategies and efficient tactics in *The 10-Minute Gardener* to help readers maximize their garden's potential while minimizing time and effort. Through her engaging YouTube channel, social media presence, and online resources, CaliKim inspires and educates gardeners of all skill levels to enjoy the rewards of homegrown bounty.

Join CaliKim on this garden adventure and discover how just a few minutes a day can yield abundant harvests and deeper connection to nature.

WHERE TO FIND CALIKIM

Seed & Garden Shop:
calikimgardenandhome.com

YouTube:
youtube.com/user/CaliKim29

Instagram:
instagram.com/calikim29

Facebook:
facebook.com/CaliKimGardenandHomeDIY